W9-CRB-516

Metals and Life

Metals and Life and *Concepts in Transition Metal Chemistry* have been written as part of the Open University course S347 Metals and Life and are designed to work as stand-alone textbooks for readers studying them either as part of an educational programme at another institution, or for self-directed study.

Details of this and other Open University courses can be obtained from the Student Registration and Enquiry Service, The Open University, PO Box 197, Milton Keynes MK7 6BJ, UK; tel: +44 (0)845 300 60 90; email: general-enquiries@open.ac.uk. Alternatively, you may wish to visit the Open University website at www.open.ac.uk, where you can learn more about the wide range of courses and packs offered at all levels by The Open University.

Metals and Life

Edited by Eleanor Crabb and Elaine Moore

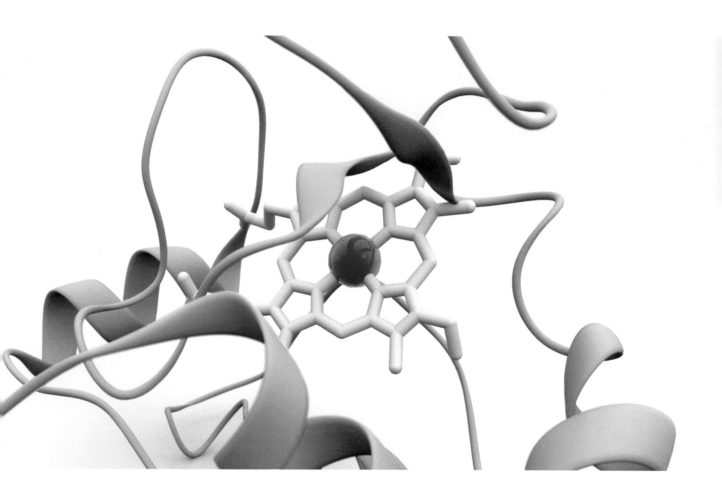

RSC Publishing

The Open University

Chem

Published by

Royal Society of Chemistry
Thomas Graham House
Science Park, Milton Road
Cambridge
CB4 0WF
United Kingdon

in association with

The Open University
Walton Hall, Milton Keynes
MK7 6AA
United Kingdom

Edited and designed by The Open University.

Typeset by The Open University.

Printed and bound in the United Kingdom by Halstan Printing Group, Amersham.

The paper used in this publication is procured from forests independently certified to the level of Forest Stewardship Council (FSC) principles and criteria. Chain of custody certification allows the tracing of this paper back to specific forest-management units (see www.fsc.org).

This book forms part of the Open University course S347 *Metals and life*. Details of this and other Open University courses can be obtained from the Student Registration and Enquiry Service, The Open University, PO Box 197, Milton Keynes MK7 6BJ, United Kingdom (tel. +44 (0)845 300 60 90, email general-enquiries@open.ac.uk).

www.open.ac.uk

British Library Cataloguing in Publication Data available on request

Library of Congress Cataloguing in Publication Data available on request

ISBN 978 1 84973 059 4 paperback

ISBN 978 1 84973 061 7 hardback

1.1

Preface

This book aims to provide an introduction to the fascinating field of bioinorganic chemistry.

We begin by introducing you to the metals essential for life, and the functions that they fulfil in the physiology of animals. These metals primarily exist as complexes and so the second chapter looks at a number of ligands of biological importance. (Prior study in inorganic chemistry, in particular coordination chemistry, is assumed, together with an appreciation of the techniques used to the characterise metal complexes and proteins.) Following this introduction, we then move on to consider the methods that organisms employ to acquire metal ions, transport and store them. This leads on to a discussion of biomineralisation, important in the formation of bone and teeth. Later chapters introduce you to the roles that metals play in biology and some of the key processes involving 'metalloproteins'. The final chapter (which is delivered online) uses excerpts from textbooks in the RSC eBook collection and embedded videos to consider the role that metals play in medicine, in the diagnosis and treatment of disease, as well as the effects of metal toxicity and deficiency.

A few words now about the layout and style of the text. At various points in the book, you will find 'boxed' material; this provides background information or enrichment materials outside the scope of the main narrative. You will also find important terms highlighted in **bold** font in the text at the point where they are first defined, and these terms are also bold in the index.

Active engagement with the material throughout the book is encouraged by the use of questions incorporated into the text, indicated by a square (■), followed immediately by our suggested answer. In addition, further questions testing your understanding of the materials are included on the website associated with this book (indicated in the text by the 💻 icon). If you are studying this book as part of an Open University course you should visit the course website. If you are not reading this book in conjunction with an Open University course of study, further resources are available from the accompanying website by visiting www.rsc.org/metalsandlife.

We would like to thank the many people who helped with the production of this book. In addition to the principal authors, Joan Mason and Kiki Warr contributed to the text. We would also like to thank the authors of the books in the RSC eBook collection that are referred to in Chapter 9.

In addition we would like to thank all those involved in the Open University production process, Margaret Careford for her careful word processing, Roger Courthold for transforming our rough sketches into colourful illustrations, Chris Hough for cover design and artwork, Hazel Carr, Yvonne Ashmore and Judith Pickering for managing the whole process and to our editor Rebecca Graham whose help in the editing of this book, is very gratefully acknowledged. We would like to thank our External Assessor, Professor Kieran Molloy, University of Bath and critical readers, Dr Ruth Durant, University of the West of Scotland, Dr Christine Gardener and Dr Susan Dewhurst, whose detailed comments have contributed to the structure and content of the book. Finally, we are delighted that this book is being published in association with RSC Publishing.

Contents

1 Introduction

At first sight the idea of inorganic chemistry associated with life may appear to be a rather narrow field of study, as we tend to think of living matter as being just organic (even suggested in the name 'organic' chemistry). However, it is a fact that without certain inorganic elements no organism could exist.

Figure 1.1 shows a 'biological Periodic Table'. The elements that are known to be essential in biochemical systems for a wide range of plants, animals or bacteria are shown in orange and blue, with those thought to be essential or possibly essential for only some species in pink and green respectively.

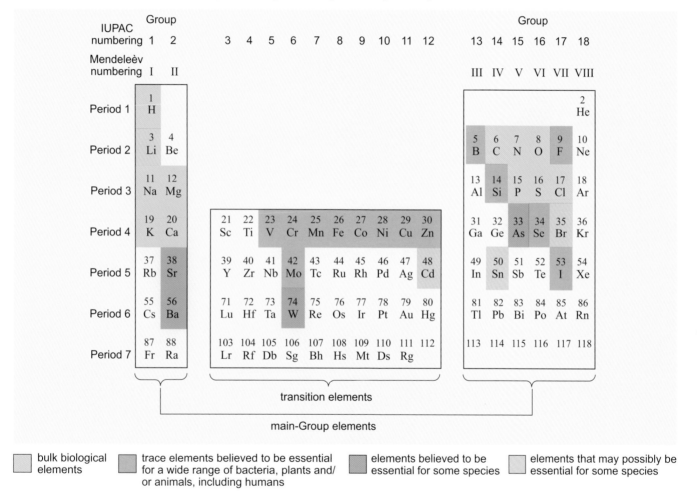

Figure 1.1 Biological Periodic Table of the elements. Elements that occur naturally and are essential for many biological systems are highlighted in blue; those believed to be essential or possibly essential for at least one species are shown in pink and green, respectively. Bulk biological elements are shown in orange.

The bulk organic elements, carbon, hydrogen and oxygen, together with nitrogen, make up 99% of the human body (shown in Table 1.1). In particular, the high percentage of hydrogen and oxygen reflects the high water content present in living systems.

Table 1.1 Percentage of atoms in the human body.

Element	Atom/%
hydrogen	62.8
oxygen	25.4
carbon	9.4
nitrogen	1.4
other	1.0

The field of bioinorganic chemistry as a research area is relatively new, having become established in only the past 50 years or so, focusing on the role of inorganic elements, particularly metals, in biological systems.

The term '**bioinorganic**', which is a composite of biology and inorganic, is used to describe the occurrence and properties of inorganic elements in living systems. Excluding nitrogen, these make up the 1.0% of 'other' elements in Table 1.1.

It is clear that these elements occur in most groups of the Periodic Table, although they are predominantly from the higher Periods and include both metals and non-metals. The wide distribution across the Periodic Table is perhaps not surprising given the wide variation of roles that these elements undertake. You may note that, as a group of elements, the transition metals are especially important.

It is not always clear whether an element is essential or not; indeed some elements are only essential for a particular species, for example tungsten appears to be present in the enzymes of *hyperthermophilic archaea*, organisms that thrive around hydrothermal vents under the sea. It should be noted however, that just because an element is present in an organism, it does not mean that it is necessarily essential. Nutrients have a metabolic route that involves mechanisms for uptake, transport, regulation, storage, utilisation and ultimately disposal. Some elements may enter the metabolic route of another (essential) metal and in some cases may interfere with biological processes of that element. For example, the average human contains approximately 300 mg strontium which chemically resembles calcium, but is not believed to be essential for our health; it is also, fortunately for us, not toxic. (Strontium is however believed to be essential for some corals.) A lack of detailed knowledge about the exact role of each element can make classification difficult.

In this book we will explore different roles that metal ions play in biological systems. We will first look at the diverse functions that some of the inorganic elements in the biological Periodic Table undertake in humans. Many of these elements will also be essential for other animals, plant life and bacteria, and although there will be some commonality, the exact functions understandably will differ in many cases.

1.1 Essential inorganic elements for human life

Table 1.2 shows the typical mass of some of the essential bioinorganic elements in a 70 kg human (collectively included as the 'other' elements in Table 1.1), together with their biological functions. Serious disorders (also listed in Table 1.2) may result from their deficiency (or indeed overload). By studying these elements in a biological context, not only can we learn what roles they play in biology, but also we can provide a foundation for understanding, and eventually treating, many of the health problems listed. The major dietary source of each element is also included.

Table 1.2 Mass and dietary sources of bioinorganic elements in a typical 70 kg human, and the functions that they undertake.

Metal	Mass/ g	Dietary source	Function	Effect of deficiency
calcium	980	dairy, vegetables, sardines	structure, charge carrier	retarded skeletal growth
phosphorus	770	fish, meat, eggs, dairy	structure, ATP	
sulfur	140	fish, meat	amino acids	
potassium	140	fruit, nuts	charge transfer, regulation of intracellular fluid	*
sodium	98	cereals, salt	charge transfer, regulation of intracellular fluid	*
chlorine	84	salt	utilisation of glucose	mild diabetes*, reduced cholesterol
magnesium	19	nuts, chocolate	structure	muscle cramps, convulsions
silicon	18	cereal	unclear[†], possibly present in connective tissue	inhibited growth
iron	4	red meat, fortified cereals	oxygen transport and storage, electron transfer	anaemia, immune system disorders
fluorine	2.6	water, meat, egg, tea, dairy	structure	inhibited growth, infertility, anaemia, dental decay
zinc	2.3	red meat, cheese, herrings	structure, enzyme	skin damage, stunted growth
copper	0.07	seafood, meat, nuts	component of many enzymes (e.g. cytochrome c oxidase)	anaemia, artery weakness
manganese	0.014	cereal products, nuts	enzyme, glucose metabolism	infertility, inhibited growth
molybdenum	0.007	meat, egg, beans	enzyme	poor cell growth
nickel	0.007	vegetables, pulses	enzyme	inhibited growth
selenium	0.0035	cereal, bread	enzyme	anaemia, infertility
chromium	0.0021	wheatgerm, kidney	unknown[†], possible involvement in glucose tolerance	diabetes symptoms
vanadium	0.0021	seafood, liver	regulates enzyme	inhibited growth
cobalt	0.0014	vitamin B_{12}, sardines, egg, liver	component of vitamin B_{12} (cobalamin)	pernicious anaemia
iodine	0.0014	milk, fish	regulates metabolic function (temperature)	goitre (swollen neck due to enlarged thyroid), retarded metabolism

* Sodium, chlorine and potassium deficiency is rare, and acute cases only tend to occur with severe dehydration.

[†] Role is not certain.

These elements can crudely be classified into either bulk or trace elements. A bulk element makes up a significant percentage by mass of most organisms (>0.1%), whereas a trace element is present in most organisms in only very small quantities.

Table 1.2 shows three clear classes of elements. The bulk metals, calcium, potassium, sodium and magnesium form one class, and together with the other bulk elements, phosphorus, sulfur, chlorine and silicon, make up just less than 1% of the atoms in the body. These elements (mostly in ionic form) occur widely throughout the body and indeed are essential for *all* life. Calcium and phosphorus (as phosphate), are particularly abundant in animals, owing to their presence in bone. Calcium is also found in a variety of proteins and enzymes, and together with sodium and potassium, it is important for signal transmission in nerves. Sulfur (and the other bulk inorganic element, nitrogen) is present in amino acids and proteins.

A second class of elements contains the trace elements including iron, fluorine, zinc and copper, which are found in small quantities and are required by most biological systems. Both iron and zinc ions are found in blood proteins, and are also components of a large number of enzymes, as is copper. Fluorine is found in bone and teeth, and is added to drinking water and toothpaste to help prevent corrosion of tooth enamel.

A third class contains elements that are found in very small amounts and includes the trace metals manganese, chromium, molybdenum, cobalt and vanadium. These are sometimes referred to as 'ultra trace' elements and in common with the other essential transition metals discussed above are often important components in enzymes.

■   How do humans maintain adequate levels of these inorganic elements in our bodies?

☐   We acquire these elements from the food and water that we consume. A list of dietary sources is given in column 3 of Table 1.2.

Of the inorganic elements, it is the important role that metals in particular play in life that will form the basis of this book. We will briefly consider some of the key functional roles that metals undertake in biological proteins in the next section. But first you should note that other than the alkali and alkaline earth metals, most metals in biological systems are associated or coordinated with ligands as **coordination compounds** or **complexes**. In Chapter 2, we will consider three major classes of 'biological' or 'biochemical' ligands that bind with metals: peptides or proteins with suitable amino acid side chains, macrocyclic ligands, such as porphyrins, and finally, the nucleobases found in DNA. The ability of the transition metals to form coordination compounds is also important in the mechanisms involved in the acquisition, transport and storage of metals as we will see in Chapters 3–5 of this book.

1.2 Functional roles of metals

As you saw in Table 1.2, many biological processes involve metals. Metals play a role at a number of different levels as illustrated in some key examples below. In many of these examples the metals are constituents of proteins, forming a group of proteins called **metalloproteins**. The metal ions, commonly described as 'metal cofactors', can help to perform physiological functions. The metal can be present as a single ion, a pair of ions or as a cluster.

Let's start, at the molecular level, with perhaps one of the most well-known examples of a metal in living systems, iron. Iron is required by almost all organisms and has an essential role to play in many enzymes and proteins. In mammals, including humans, iron is an integral part of the blood; a deficiency leads to anaemia, a condition that affects many people. Accordingly, the bioinorganic chemistry of iron has been studied extensively, with many studies concentrating on the blood protein **haemoglobin**, which is an oxygen-carrier in the blood. The iron centre of haemoglobin, called **haem**, shown in Figure 1.2a, provides a site to which the oxygen molecule can bind. The iron in haem, acts as a **Lewis acid**, accepting a non-bonding or lone pair of electrons from the oxygen. This binding is reversible, allowing transport and delivery of oxygen to where it is required.

You will recall from Figure 1.1 that the transition metals are a particularly important group of bioinorganic elements (many of the trace metals in particular are in this group). One important feature of the transition metals is their ability to exist in more than one oxidation state. Iron, for example, is found typically as either the Fe^{2+} or Fe^{3+} ion. This ability of iron and the other transition metals to exist in variable oxidation states is at the root of their ability to function as **electron transfer agents**, transferring electrons to enzymes to perform a specific function. One example is the 4Fe–4S cluster of ferredoxin illustrated in Figure 1.2b, important in bacteria for nitrogen fixation, which can take an overall charge of −3, −2 or −1 depending on the charge on the iron.

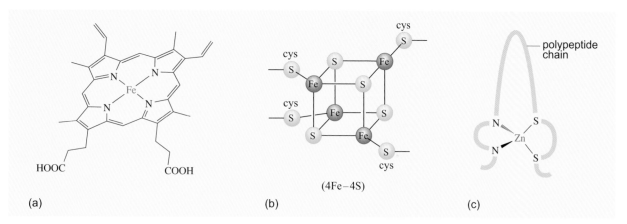

(a) (b) (4Fe–4S) (c)

Figure 1.2   Structure of (a) haem, (b) 4Fe–4S cluster in ferredoxin and (c) Zn finger. Cys is the amino acid cysteine.

Metal ions may also be incorporated (deeply buried) into metalloproteins in a structural role. An example is that of Zn^{2+} metal ions which are found in many proteins, including 'zinc-finger proteins' (Figure 1.2c) which recognise and bind to particular sequences of DNA (deoxyribonucleic acid) and are important in transcription. The Zn^{2+} ion is believed to be essential in maintaining the protein's three-dimensional structure. On genetic evidence, it has been suggested that there may be up to 200 zinc-finger proteins waiting to be characterised.

Metals are also present in a subclass of metalloproteins, called **metalloenzymes**. (Enzymes are proteins that act as biological catalysts.) Many of the physiological reactions important in life are catalysed by metalloenzymes. Examples include the catalytic reduction of O_2 to water (important in respiration) and the reduction of nitrogen to ammonia (important in nitrogen fixation in certain groups of bacteria). In these metalloenzymes the metal ion may be part of the catalytic active site itself or used to transfer atoms or groups to the active site.

Note that the relationship between the metal and protein can work in two ways. The presence of the metal can influence the electronic and structural features of the protein. The protein can also influence the properties of the metal, for example by stabilising unusual coordination geometries or particular oxidation states. The complex interplay between these two aspects will determine the reactivity in the system. We will come across many examples of this throughout the book.

Metal ions can also have a role on their own as counter ions. For example Mg^{2+} ions have a structural role, stabilising the DNA double helix by binding the phosphate groups so preventing the strands from repelling each other.

At the cellular level, metals ions, in particular the alkali and alkaline-earth metal ions Na^+, K^+ and Ca^{2+}, are used in biology in communication roles to trigger cellular response. For example, a rapid influx of Na^+ ions into a cell triggers the firing of neurons in nerves.

Solid calcium compounds also play a structural role at the macroscopic level, as a major component in bone, teeth and shell. Metal ions, in particular iron in magnetite, Fe_3O_4, may be used by some organisms as internal magnetic compasses by which to navigate. Magnetotactic bacteria, for example, use magnetite to orient themselves along the Earth's magnetic field lines.

This section is just a brief taster of some of the chemistry which influences the functions that metals can play in biological processes. These will be developed further throughout this book, in particular in Chapters 6–8.

■ List the various functions of the metal in the examples above.

☐ In the examples discussed above, the metal ions in metalloproteins are important for oxygen transport, electron transfer and also in structural roles. Metal ions are also important components of metalloenzymes, either at the active site, or responsible for delivering reactants to the active site. Metal ions can also have a communication role, and may also provide structure at a macroscopic level, in bone, for example.

■ What key properties do transition metals display that are utilised in metalloproteins?

☐ The two key properties, illustrated in the examples above, are the ability of transition metals to occur in different oxidation states and also the Lewis acid behaviour of metals.

Finally, before moving on you should note that, in addition to these naturally occurring inorganic elements, metals are also introduced into biological systems for medicinal use with widespread application in both diagnostics and therapy (Table 1.3). Examples such as these will be considered in the online chapter (Chapter 9) of this book.

Table 1.3 Examples of metals used in medicine and their application.

Metal	Medicinal use
Pt, Ru	therapy: anticancer therapy
Ba	diagnostic: Ba meals in X-ray
Gd	diagnostic: MRI contrast agent
Tc	diagnostic: radiopharmaceutical for functional imaging
Re	therapy: radionuclide for therapy
Au	therapy: arthritic drug
V	therapy: diabetes (insulin mimic)
Bi	therapy: treatment for peptic ulcers
Li	therapy: drug for bipolar disorders

1.3 Bioavailability of metals

As you have already seen, certain metals are essential for life. Remarkably, there is very little variation in the relative proportion of the elements from one person to another. Indeed, any significant variation can lead to disease.

One question we might ask is why has an organism chosen one specific element rather than another. A major factor will be the chemical suitability of the element for a particular function, as we have seen briefly in the previous section, but it will also depend on their availability (both now and in the past).

For an element to be of use to an organism, it must be 'available' in the local environment of the organism. This will depend on two factors. First, the abundance of the element will be important. Second, the element must be in an accessible or extractable form such that it can be taken up or acquired by an organism. This availability of metals to biological organisms is referred to as **bioavailability**.

Although we acquire most of our metal nutrients from our diet, this will ultimately depend on the absorption of these nutrients by plants from the soil. The abundance of the different metals in the Earth's crust is shown in Table 1.4. Almost all metals occur as oxide, carbonate or sulfide ores, as shown in column 3 of the table.

Table 1.4   Elemental composition of the Earth's crust and seawater for the essential metals.

Element	Earth's crust/ppm	Major ores	Seawater/ppm
Ca	5×10^4	$CaCO_3$, $CaMg(CO_3)_2$, $CaSO_4.2H_2O$, CaF_2	4×10^2
Co	30	CoS_2, $CoAsS$, $CoAs_2$, Co_3S_4	$< 1 \times 10^{-4}$
Cu	68	$CuFeS_2$	3×10^{-3}
Fe	6×10^4	Fe_3O_4, Fe_2O_3, $FeO(OH)$	3×10^{-3}
K	1.5×10^4	KCl	416
Mg	3×10^4	$CaMg(CO_3)_2$, $MgCO_3$	1.3×10^3
Mn	1×10^3	$MnCO_3$, MnO_2, $MnO(OH)$	2×10^{-3}
Mo	1.1	MoS_2	0.01
Na	2.3×10^4	$NaCl$, $Na_3(CO_3)(HCO_3)$	1.1×10^4
Ni	90	$(Fe,Ni)_9S_8$	2×10^{-3}
V	190	$K(UO_2)(VO_4).1.5H_2O$, $PbCl_2.3Pb_3(VO_4)_2$	1.5×10^{-3}
Zn	79	ZnS	5×10^{-3}

■ According to Table 1.4, which are the two most abundant essential metals in the Earths crust?

☐ Calcium and iron appear to be the most abundant metals. As we will see in the next section, the abundance of a metal in the Earth's crust however does not necessarily reflect its true availability.

1.3.1   Aqueous chemistry

Much of the chemistry of life occurs in aqueous media, highlighted by the high water content of organisms (emphasised by the percentages of H and O in the human body in Table 1.1). Most of the metals in our diet are in the form of water-soluble salts. Thus the solubility of metals is an important factor to consider. Solubility will depend on the nature of the metal salt, and also on the pH and temperature of the solution. Human life is restricted to a narrow range of conditions, generally a pH of 7 and a temperature of 303 K.

As the concentrations of metals in solution are likely to be more important generally than the amount of an element present in rocks (i.e. solid compounds), we are justified in considering that the availability of elements may be better reflected if we consider the abundance of the elements in the oceans, shown in Table 1.4 (column 4). When we compare the abundance of the elements in the oceans with their overall abundance in the Earth's crust, quite a different pattern emerges.

For example, the iron content of the Earth's continental crust is actually relatively high (4.1%), but most of it exists as the highly insoluble – and, therefore, unassimilable (not able to be incorporated into) – compounds, hydrated iron oxide (hematite), $Fe_2O_3.nH_2O$, iron hydroxide, $Fe(OH)_3$, magnetite, Fe_3O_4, or siderite, $FeCO_3$. This is reflected in the low concentration of iron in seawater. This however has not always been the case (Box 1.1).

Box 1.1 The primitive sulfide sea

The concentration of elements in the sea has not always been as it is now. For example, iron in ancient seawater is estimated to be up to 1000 times more concentrated than it is today. Its concentration is related to the atmospheric oxygen content. Until around two billion years ago, the Earth's atmosphere was anoxic (low free oxygen content), believed to consist of several gases including CO/CO_2, N_2, H_2, CH_4 and H_2S. Consequently, in the seas at this time the elements were typically present as the sulfides, and iron was present as the more soluble iron(II) sulfide (Figure 1.3). As the level of oxygen present in the atmosphere increased (from photosynthesis as algae and plants evolved), solubility depended largely on the formation of hydroxides and carbonates, rather than sulfides. Iron was precipitated as the highly insoluble iron(III) hydroxide.

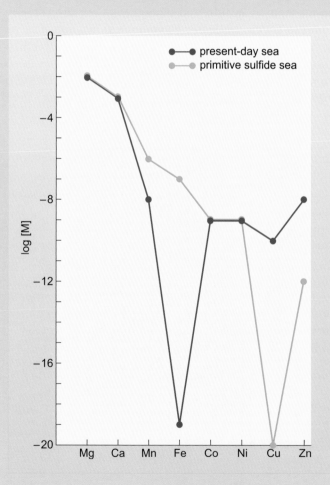

Figure 1.3 The concentration of free elements in the sea: (light blue) in the primitive sulfide sea; (dark blue) in the aerated sea of today. Note that the concentrations are expressed on a logarithmic scale.

■ How might you account for the very low concentration of copper in the primitive sea?

> □ Copper sulfide is particularly insoluble compared to the carbonate or hydroxide, accounting for the higher concentration seen in the present-day oceans.

We will explore the idea of bioavailability further in Chapter 3, and discuss some of the mechanisms that organisms can employ to *selectively* take up the elements they need, even when the elements are present in only very small amounts.

1.4   How much of a good thing?

Let us now take a short aside and consider a hereditary (i.e. genetic) human disease, **thalassaemia**, which affects levels of iron within the body. Thalassaemia patients suffer from similar symptoms to anaemic patients, as essentially both have low counts of healthy red blood cells. The disease is called thalassaemia (from the Greek: *thalassa*, meaning sea) because it is very common in countries surrounding the Mediterranean Sea. It is in addition widespread in Central Africa, India and South East Asia. Thalassaemia is also known as Cooley's anaemia, after the American physician who first identified it. The disease takes two forms: thalassaemia-minor and thalassaemia-major. The former usually has symptoms of mild anaemia, whereas the latter is seriously debilitating.

The treatment for thalassaemia-major is regular blood transfusions, which restore the healthy red-blood cell count to normal levels. However, this treatment can lead to its own problems. The problems are not directly linked to the symptoms of anaemia, which are largely ameliorated by the transfusions, but to the regular intake of large quantities of iron into the body. The human body is capable of excreting a maximum of about 10 mg of iron a day, whereas regular blood transfusions put far more iron into the body. This leads to a condition known as iron overload or **haemochromatosis**. One of the symptoms of iron overload is increased susceptibility to bacterial infection. The extra iron in the body can be absorbed by other organisms, particularly bacteria (remember that not only humans require iron to live), which then proliferate and cause infection.

Owing to its high mortality rate and widespread occurrence, thalassaemia is the subject of much medical research. It does, however, highlight some important principles for the bioinorganic chemist. First, it is clear that within the healthy human body, biochemical systems carefully control the level of iron so that it is around an optimal value (this control is known as **homeostasis**). Second, it appears as if the human body has a means of transporting iron and storing it. For example, iron stores are required in case of sudden loss of iron, say from heavy bleeding. It is also important during transport and storage that iron is not made available to other organisms, thus helping to reduce the possibility of bacterial infection. We will consider the transport and storage of iron in detail in Chapters 4 and 5.

Iron serves as a good example of the delicate balance between beneficial and toxic levels of intake. Too little iron can lead to anaemia; too much increases a body's susceptibility to bacterial infection. From the illustration above, it is clear that there is an optimal intake level of iron, just as there is for all bioinorganic elements, in the form shown schematically in a **Bertrand diagram** in Figure 1.4. The overall shape of the curve in Figure 1.4 is common to all the bioinorganic elements. The position and broadness of the curve differ widely from element to element and from organism to organism. For example, mercury is toxic to humans in very low concentrations. (The term 'mad-as-a-hatter' originally referred to workers in the hat industry who used $Hg(NO_3)_2$ to polish hat felts. Long term, low-level exposure to mercury is now known to lead to dementia.) Despite the high toxicity of mercury to humans, some microorganisms however have a much greater tolerance of high mercury concentrations. Unlike humans, these remarkable microorganisms are able to dispose of ingested mercury by biochemically converting it into dimethylmercury, $[Hg(CH_3)_2]$. Dimethylmercury is relatively volatile (with a boiling temperature of 93 °C), and simply evaporates from the microorganism. As a result, these microorganisms are able to exist in mercury-contaminated areas. The mercury state-of-health curve for the microorganisms is thus much broader than the same curve for humans.

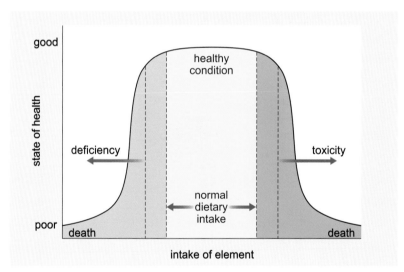

Figure 1.4   Schematic diagram of 'state of health' versus intake concentration of a particular element.

1.5   Protein structure

Many of the examples we have met so far have been metalloproteins, where a metal ion is bound via amino acid groups to a protein. We will now take time to consider the structure of proteins. The function of a protein will depend on its precise three-dimensional structure, which in turn depends on the sequence of its constituent amino acid monomers.

The 20 commonly occurring amino acids found in proteins in bioinorganic chemistry are shown in Table 1.5.

Table 1.5 The 20 naturally occurring α-amino acids; most amino acids have the general formula H_2N–CHR–COOH; the side chain R may have any of the structures shown here.

Amino acid	Symbol	One-letter code	Side chain (R group)	Amino acid	Symbol	One-letter code	Side chain (R group)
alanine	Ala	A	—CH_3	lysine	Lys	K	—$(CH_2)_4$—NH_2
arginine	Arg	R	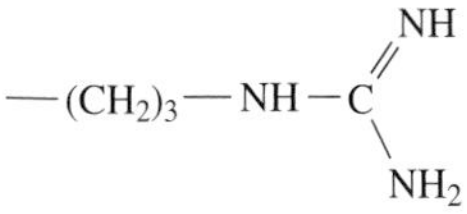 —$(CH_2)_3$—NH—C(=NH)(NH_2)	methionine	Met	M	—$(CH_2)_2$—S—CH_3
asparagine	Asn	N	—CH_2—$CONH_2$	phenylalanine	Phe	F	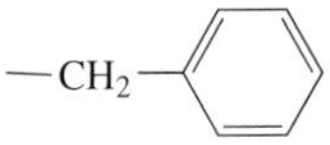 —CH_2—C$_6$H$_5$
aspartic acid	Asp	D	—CH_2—COOH	proline (whole structure)	Pro	P	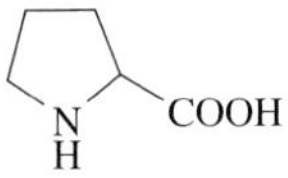 (pyrrolidine ring with COOH)
cysteine	Cys	C	—CH_2—SH	serine	Ser	S	—CH_2—OH
glutamate	Glu	E	—$(CH_2)_2$—COOH	threonine	Thr	T	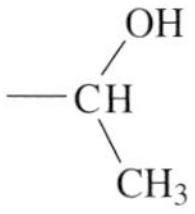 —CH(OH)(CH_3)
glutamine	Gln	Q	—$(CH_2)_2$—$CONH_2$	tryptophan	Trp	W	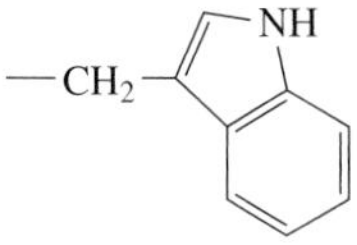 —CH_2—(indole)
glycine	Gly	G	—H	tyrosine	Tyr	Y	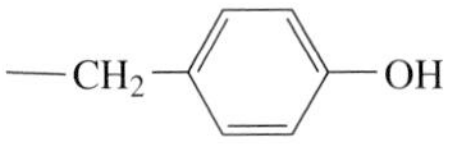 —CH_2—C$_6$H$_4$—OH
histidine	His	H	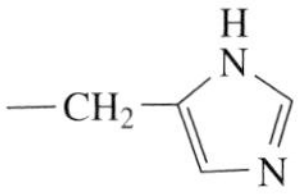 —CH_2—(imidazole)	valine	Val	V	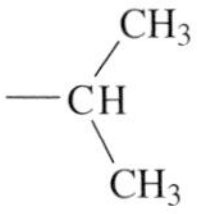 —CH(CH_3)$_2$
isoleucine	Ile	I	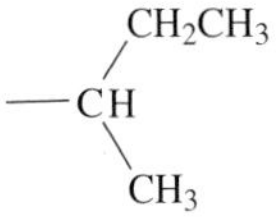 —CH(CH_2CH_3)(CH_3)				
leucine	Leu	L	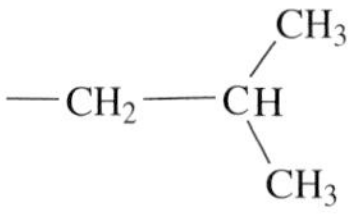 —CH_2—CH(CH_3)$_2$				

As we shall see when we look at biological ligands in Chapter 2, a characteristic feature of the amino acid side chains is an ability to coordinate to a metal ion via a lone pair of electrons on an electronegative atom such as oxygen, nitrogen and sulfur. Now, however, we will concentrate on the general structure of proteins and how they are determined.

Typical proteins have relative molecular masses in the range of 15 000–70 000, and contain typically 140–640 amino acids. Variations in the number and sequences of the amino acids give rise to the huge diversity of proteins and their structures. This is particularly true for the globular proteins, which include many enzymes. In contrast, fibrous proteins, such as collagen, which are long and thin, are regular repeats of just a few particular amino acids and so their structure is more uniform. The complexity of protein structure, summarised in Figure 1.5, is described in terms of a four-tier hierarchy, from primary, through three levels of higher-order structure (secondary, tertiary and quaternary).

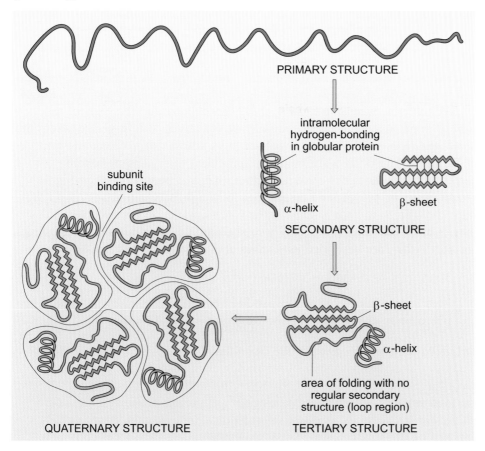

Figure 1.5   The structural hierarchy of proteins.

1.5.1   Primary and secondary structure

The **primary structure** of a protein is simply the sequence of amino acid monomer units, or residues, of which it is composed.

A single polypeptide chain is not rigid. Rotation about each of the α-carbon atoms of the amino acid residues is possible (except proline), which allows for a very flexible structure and folding of the chain in a theoretically huge number of ways. The folding of the linear polypeptide chain into its precise higher-order structure is largely determined by its amino acid sequence. In general terms, the chain folds into the most energetically favourable shape, or **conformation**; in this conformation, hydrophobic ('water-hating') amino acid side chains are clustered together in the interior of the molecule, away from the aqueous environment of the cell, and most hydrophilic ('water-loving') side chains are on the surface of the molecule where they can interact with surrounding water molecules. There are, however, regions of so-called **secondary structure** in which the polypeptide chain is folded into regular, stable patterns.

■ What type of interaction is possible between the N–H group of one amino acid and the C=O group of another?

☐ As shown in Figure 1.6, the formation of hydrogen bonds is possible between these groups. These hydrogen bonds are responsible for the secondary structure of the protein.

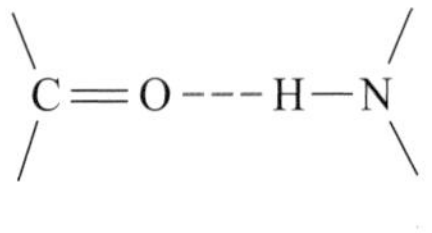

The two most common folding patterns, shown in Figure 1.7, are the **α-helix** and the **β-sheet**. In the α-helix, there is hydrogen-bonding in the direction of the helix axis between the C=O of one peptide bond and the N–H of the peptide bond four amino acids units along the chain. This structure places the polar, hydrogen-bonded C=O plus N–H pairs within the hydrophobic interior of the helix. The β-sheet structures are produced by hydrogen-bonding perpendicular to adjacent chains, via their opposed peptide C=O and N–H groups. Pairs of polypeptide chains that form a β-sheet can either run in the same direction (i.e. both N-terminal to C-terminal) or in opposite directions (one N to C, the other C to N). These alternative β-sheet arrangements, which are both stable, are described as parallel and antiparallel, respectively.

■ What type of β-sheet is illustrated in Figure 1.7?

☐ This is an example of an antiparallel β-sheet, as the adjacent parts of the polypeptide chains run in opposite (antiparallel) directions.

In fibrous proteins the polypeptide chain tends to be folded into a uniform secondary structure. Globular proteins, however, do not have a regular secondary structure throughout. Between the sections of α-helix and β-sheets are **loop regions**, which have an irregular yet precise shape (Figure 1.5), giving the molecule its unique conformation. These sequences are usually exposed on the surface of a protein molecule, where they often contribute directly to its specific functions. (We will meet an example of this in Chapter 4 in ion channels.)

1.5.2 Tertiary structure

The **tertiary structure** of a protein is the three-dimensional arrangement of the entire polypeptide chain: α-helix, β-sheet and all the intervening loop regions. Globular proteins are compact and roughly spherical in shape, whereas fibrous proteins tend to be long and thin.

■ What types of interaction between R groups are likely to be important in maintaining tertiary structure?

☐ Hydrogen-bonding, for example between the O–H group of one amino acid and the C=O group of another, ionic bonding between charged species and van der Waals forces. As discussed above, packing of hydrophobic amino acids together, usually within the interior of a protein

$$\begin{array}{c}\diagdown \qquad\qquad \diagup \\ C{=}O\text{---}H{-}N \\ \diagup \qquad\qquad \diagdown\end{array}$$

Figure 1.6
Hydrogen bond formation in protein secondary structure between the N–H group of one amino acid and the C=O group of another.

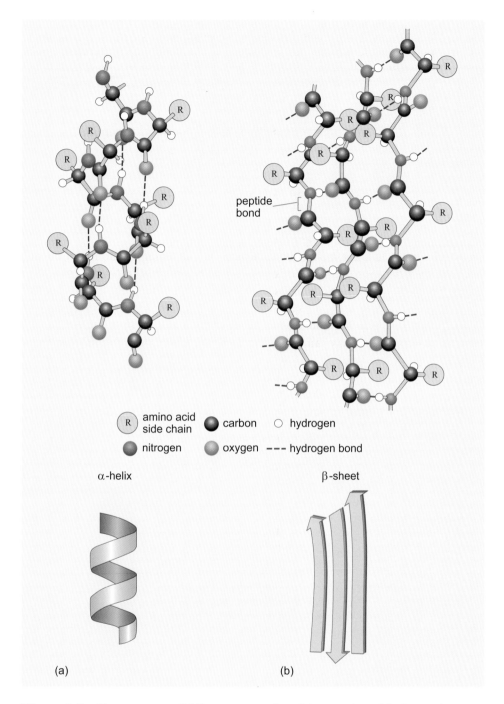

peptide bond

R amino acid side chain   ● carbon   ○ hydrogen
● nitrogen   ● oxygen   --- hydrogen bond

α-helix   β-sheet

(a)   (b)

Figure 1.7 Two common folding patterns found in proteins; (a) shows the structural and schematic (ribbon) representations of the α-helix, respectively, and (b) those of a β-sheet. The side chains are simply denoted as R and α-helices are shown in red and β-sheets in blue.

to avoid interaction with external polar solvent, may also be important. In addition, covalent bonding is possible such as in a **disulfide S–S bridge** between two cysteine residues. These are summarised in Figure 1.8.

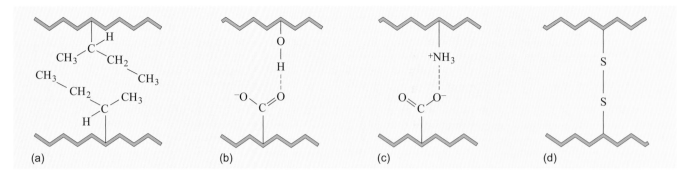

Figure 1.8 Types of interaction between amino acid side chains. (a) Van der Waals forces between hydrocarbon groups, (b) one example of hydrogen bonding, (c) ionic bonds between oppositely charged groups, and (d) a disulfide S–S bridge.

As well as interactions between the residues, there may also be interactions with a metal or other cofactor and/or with the environment such as water molecules. The tertiary structure of a protein is significant for its biological activity. Many proteins, other than those with a structural role, function by interacting with specific molecules or ions. These tend to bind at specific binding sites. Different proteins will bind to a vast range of ions from small molecules, or other proteins, through to the largest molecules of all, DNA. The important functional point is that this binding is highly specific, such that a given protein may bind to only one or two species. This is an important concept and we shall return to this throughout the rest of this book.

1.5.3 Quaternary structure

For globular proteins that are made up of more than one polypeptide chain, there is a further tier in the structural hierarchy called the **quaternary structure**. Here, the constituent polypeptide chains, each folded into its tertiary structure, are held together by weak, non-covalent interactions. In this context, the polypeptide chains are called subunits. Figure 1.9 shows the quaternary structure of haemoglobin.

■ How many subunits make up haemoglobin? What other features in the structure of haemoglobin are evident?

□ The quaternary structure of haemoglobin consists of four subunits. Each subunit is a single polypeptide chain with a haem group. Notice that the subunits of haemoglobin are non-identical; there are two α- and two β-subunits. We will meet these again in Chapter 8.

1.6 Biomolecule study

Much of our current knowledge of bioinorganic chemistry comes from studies of inorganic elements in biological systems. A range of physical techniques is available for determining the structure and other properties of proteins and their metal centres listed in Table 1.6. Of these techniques, single-crystal X-ray

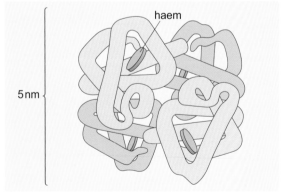

Figure 1.9 The quaternary structure of haemoglobin. The haem groups (shown in red; Figure 1.2a) are non-protein planar structures with an iron atom at the centre where the oxygen is bound. The four polypeptide chains are shown in blue or green.

diffraction or X-ray crystallography is perhaps the most revealing, in that it gives an accurate picture of the three-dimensional structure of a protein; however, it requires growth of a crystal which is not always possible. Structural information about proteins is essential for a fuller understanding of the way proteins and enzymes function. Structures are often determined for both the metalloprotein and its **apo**-form (without the metal cofactor). In some instances, models of particular proteins may be studied; these are simpler compounds prepared to reproduce features of the structure or properties of the original metalloprotein. These are often used to give insights into the mechanism of a biochemical reaction.

Table 1.6 Techniques used to characterise bioinorganic systems of biological interest and the information that they can give.

Technique	Information
X-ray diffraction (crystallography)	determination of 3D solid state molecular structure of protein
X-ray absorption spectroscopy (XAS):	
extended X-ray absorption fine structure (EXAFS)	metal-coordination composition and geometry (metal–metal, metal–ligand distances and numbers)
X-ray absorption near-edge structure (XANES)	oxidation state of metal ion, coordination environment
nuclear magnetic resonance (NMR)	determination of 3D structure of (small) protein in solution
electron spin resonance (ESR) (also referred to as electron paramagnetic resonance (EPR))	metal-coordination composition and geometry, spin states, electronic structure
Mössbauer	oxidation state, spin state and magnetic coupling of iron
magnetic moment	oxidation state, spin state, electronic structure
electronic, UV/visible spectroscopy	structure and electronic properties, map electron transfer path (energies and intensities of electronic transitions)
vibrational spectroscopy – resonance Raman	structural information about ligand coordination to metal ion

Don't forget that there are questions on the companion website which you can use to test your understanding of the material covered in this chapter.

2 Ligands with biological significance

Many of the important metals in living systems are associated or coordinated with ligands. The metals and their surrounding ligands can be considered to be coordination compounds or complexes and the basic principles of coordination chemistry apply. In this chapter we will consider three major classes of 'biological' or 'biochemical' ligands that bind with metals in living systems. These are peptides or proteins with suitable amino acid side chains, macrocyclic chelating ligands, such as porphyrins, and finally, the nucleobases found in DNA. Other molecules such as saccharides, vitamins and hormones may also act as ligands, although we shall not consider these here. In addition, we find that certain metals are sometimes found with highly specialised ligands that have evolved to fulfil particular roles in life; we shall examine some of these metal ions in later chapters.

2.1 Amino acids

The most common biochemical ligands are amino acids, connected via peptide links into polypeptides and proteins. Amino acid side chains are often found at the active sites of metal-containing proteins (metalloproteins). Not all of the amino acids listed in Table 1.5 are found to be associated with metal ions. There are several, however, which appear to be particularly well suited for metal ion binding and these are shown in Figure 2.1. The characteristic feature of these amino acid side chains is an ability to coordinate to a metal ion via a lone pair of electrons on an electronegative atom.

■ Through which atom would you expect histidine (**2.1**) to bond to a metal ion?

☐ Histidine has a five-membered ring as part of its side chain. In its neutral form, one of the nitrogen atoms in the ring has a lone pair that can form a dative bond to a metal ion.

Note that the neutral histidine can be **protonated**; in its protonated form it is unable to coordinate to a metal ion using a nitrogen lone pair, and it will not form stable complexes with metal ions. The side chain of neutral histidine can also be **deprotonated** to give a histidate anion.

■ How does deprotonation of histidine to form the histidate anion affect its coordination properties?

☐ Its side chain can now coordinate to two separate metal ions simultaneously.

In a similar fashion, glutamate (**2.2**) and aspartate (**2.3**) are the deprotonated forms of glutamic acid and aspartic acid. The deprotonated (carboxylate) forms can act as ligands forming stable complexes with many metal ions.

■ Would you also expect that the protonated forms, glutamic acid and aspartic acid, would form complexes with metal ions? If so, how?

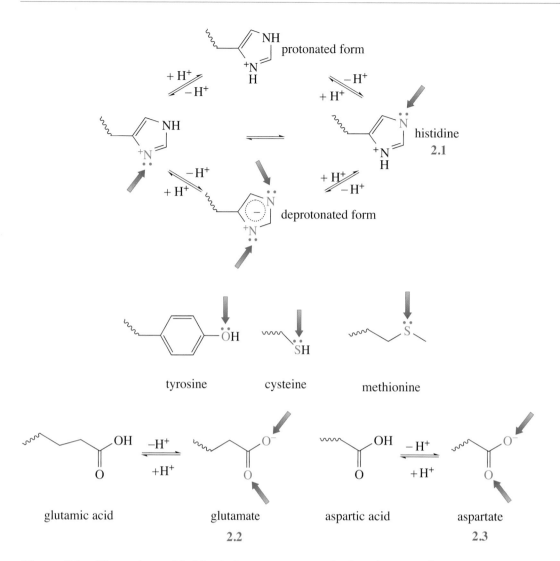

Figure 2.1   The amino acid side chains most commonly found to coordinate to metal ions in biochemical molecules. The green arrows and atoms coloured in red denote the position of coordination to a metal ion.

☐   Yes. They can coordinate using the lone pairs on the carbonyl oxygen. These complexes are not as stable as the carboxylate–metal ion complexes. The reason for this is that there is not the same extent of electrostatic attraction between the acid form and the metal ion as with the carboxylate form and metal ion.

Therefore, the protonation state of the amino acid side chain is an important factor in determining whether the side chain will form stable complexes with metal ions. A measure of the extent of protonation (at a given pH and temperature in aqueous solution) of a particular chemical group is given by the **acid dissociation constant, K_a**, and the **pK_a** of the acid form.

2.1.1   The acid dissociation constant, K_a

If we denote the protonated (acidic) form of the side chain as AH and the deprotonated (basic) form as B, then we can write an equation for the acid–base equilibrium:

$$AH(aq) = B(aq) + H^+(aq) \qquad\qquad (2.1)$$

Strictly speaking, the acid dissociation constant is defined in terms of the following equilibrium:

$$AH(aq) + H_2O(l) = B(aq) + H_3O^+(aq) \qquad\qquad (2.2)$$

which takes account of the fact that free protons, H^+, do not exist in aqueous solution. The ensuing analysis is based on the simpler Equation 2.1, but comes to the same conclusions as if we had used the more accurate Equation 2.2.

■   Write an expression for K_a for the reaction shown in Equation 2.1.

☐   The acid dissociation constant, K_a, is simply the equilibrium constant for this dissociation reaction:

$$K_a = \frac{[B(aq)][H^+(aq)]}{[AH(aq)]} \qquad\qquad (2.3)$$

where [X(aq)] is used to denote the concentration of a species X(aq).

We use K_a, and also the related quantity pK_a, where $pK_a = -\log K_a$, to determine the extent of protonation of an amino acid side chain.

The approximate pK_a values of the protonated forms of bioinorganically important amino acid side chains are given in Table 2.1.

Table 2.1   Amino acid side chains with pK_a values.

Amino acid side chain	pK_a
protonated histidine	6.5
histidine	14
aspartic acid	4.5
glutamic acid	4.5
tyrosine	10
cysteine	8.5

2.1.2 pK_a and pH

By rearranging Expression 2.3, we can see that the ratio of the concentration of base to the concentration of acid form,

$$\frac{[B(aq)]}{[AH(aq)]}, \text{ is given by } \frac{K_a}{[H^+(aq)]}$$

So, to determine the ratio of base to acid in any aqueous solution, we need to know both K_a and the acidity, which is given by the pH. The pH of most biochemical systems tends to be around pH 7.

■   How do you calculate $[H^+(aq)]$ at pH 7?

☐   pH = $-\log [H^+(aq)]$ = 7, and thus $[H^+(aq)] = 10^{-7}$ mol dm^{-3}

Let's take the deprotonation reaction of neutral histidine as a first example. We see from Table 2.1 that the pK_a is 14. Notice that the pK_a value is *higher* than the pH of most biochemical systems (around pH 7). We shall come back to this later.

■   What is the value of K_a for the deprotonation of histidine?

☐ $pK_a = -\log K_a$, so $K_a = 10^{-14}$ mol dm^{-3}

The base/acid ratio for histidine at pH 7 is therefore $10^{-14}/10^{-7}$ or 10^{-7}, and so very little histidine is deprotonated at pH 7.

Now let's take aspartic acid as an example (glutamic acid is similar). In this case the pK_a value is 4.5. Notice that the pK_a is now lower than the pH of most biochemical systems.

■   What is the ratio of the concentrations of aspartate to aspartic acid?

☐   The ratio is given by $10^{-4.5}/10^{-7} = 10^{2.5}$.

In this case we see that the base form, aspartate, dominates by a factor of about 300 (316).

These two examples demonstrate a general point: if we know the pH of the system – and we can assume pH 7 for biochemical systems – then the pK_a value for the acid–base equilibrium reaction of the side chain immediately tells us which form of the side chain dominates. So, if the pK_a is greater than 7, the protonated form dominates, and if the pK_a is less than 7, the deprotonated form dominates. Thus, Table 2.1 shows that at pH 7, *in the absence of metal ions*, aspartic acid and glutamic acid exist *mostly* as aspartate and glutamate.

Another question to ask is, by how much is the concentration of one form greater than the other? At pH 7, aspartate and glutamate as we have seen are $10^{2.5}$ (or 316) times more abundant than the protonated forms. But what about the protonated form of histidine compared with neutral histidine (this system has a pK_a of 6.5)?

■ Work out the ratio of His(aq) to HisH$^+$(aq) at pH 7. (Assume that ionisation to the histidate anion is small enough to be ignored at pH 7.)

□ If we write the equilibrium:

$$\text{HisH}^+(\text{aq}) \rightleftharpoons \text{His}(\text{aq}) + \text{H}^+(\text{aq})$$

(2.4)

$$K_a = \frac{[\text{His}(\text{aq})][\text{H}^+(\text{aq})]}{[\text{HisH}^+(\text{aq})]}$$

(2.5)

and therefore,

$$\frac{[\text{His}(\text{aq})]}{[\text{HisH}^+(\text{aq})]} = \frac{K_a}{[\text{H}^+(\text{aq})]}$$

(2.6)

$pK_a = 6.5$, so $K_a = 10^{-6.5}$, giving

$$\frac{[\text{His}(\text{aq})]}{[\text{HisH}^+(\text{aq})]} = \frac{10^{-6.5}}{10^{-7}} = 10^{0.5} = 3.16$$

(2.7)

The neutral form of histidine dominates, but now only by a factor of about three times; in other words the solution contains about 75% histidine and 25% of the protonated histidine. So histidine exists mostly in its neutral form at pH 7. Therefore, we can expect free histidine to be neutral, and aspartic and glutamic acids to be deprotonated at pH 7. The pK_a values of cysteine and tyrosine (8.5 and 10, respectively) indicate that these amino acid side chains will be protonated at pH 7.

2.1.3 The effect of metal ions

We need to remember that the pK_a values as listed in Table 2.1 are measured *in the absence of metal ions*, which of course can interact strongly with the amino acids. In neutral aqueous solution, metal ions may displace the acidic proton of an amino acid side chain, even in amino acids with pK_a values higher than 7. In these cases, we have a different equilibrium reaction, in which a side chain such as cysteinyl, which we can denote as RSH, reacts with a metal ion M^{n+}. So, it is possible to observe deprotonated cysteine coordinated to a metal ion as:

$$\text{RSH} + \text{M}^{n+} = [\text{RSH–M}]^{n+} = [\text{RS–M}]^{(n-1)+} + \text{H}^+$$

(2.8)

The tyrosine side chain, with a relatively high pK_a, is less likely to have its proton displaced by a metal ion; even so, tyrosyl side chains can coordinate to a metal in both protonated and deprotonated forms as in the example of cysteinyl above. Despite its particularly high pK_a, it is also possible to observe a deprotonated histidyl side chain coordinated to two metal ions; that is, the proton has been displaced by a second metal ion.

2.1.4 Can amide links coordinate?

The amide link, R–CO–NH–R', in a polypeptide chain does not often coordinate to metals in bioinorganic chemistry and can be ignored as a coordinating group for our purposes.

One of the reasons why the amide group is relatively poor at coordinating to metal ions can be seen from structures **2.4** and **2.5**, which are the two important resonance forms of the amide group:

2.4   **2.5**

In form **2.5** the nitrogen lone pair electrons are delocalised over the amide group, and are thus 'unavailable' for coordination to a metal ion. Even so, the oxygen atom would still be available for coordination; amide oxygens are, however, only rarely found to be coordinating atoms in bioinorganic chemistry. Further, the pK_a of the amide group is approximately 17.

■   What is the consequence of the high pK_a of the amide group?

☐   Ionisation of the proton requires a lot of energy, and so it is unlikely that a metal ion could displace the proton to form a deprotonated amide–metal ion complex.

2.1.5 Selectivity in metal binding: hard/soft acid–base theory

Metal ions and ligands can be classified as hard and soft acids and bases as in Table 2.2. (The acids and bases referred to are Lewis acids (electron-pair acceptors) and Lewis bases (electron-pair donors).) Table 2.2 also classifies the more common amino acid side chains according to **hard/soft acid–base theory**. This theory can be a useful predictor of complex stability because of the simple observation that *hard acids tend to bind to hard bases and soft acids to soft bases*. We can see that amino acids cover a wide range of hard to soft ligands. For example, cysteine, which coordinates through sulfur, can be classified as a soft ligand.

Table 2.2   Hard and soft acids and bases.

	Hard	Borderline	Soft
Acids	H^+, Li^+, Na^+, K^+, Be^{2+}, Mg^{2+}, Ca^{2+}, BF_3, BCl_3, $B(OR)_3$, Al^{3+}, $AlCl_3$, $Al(CH_3)_3$, Sc^{3+}, Ti^{4+}, VO^{2+}, Cr^{3+}, Fe^{3+}, Co^{3+}, La^{3+}	Fe^{2+}, Co^{2+}, Ni^{2+}, Cu^{2+}, Zn^{2+}, Rh^{3+}, $B(CH_3)_3$, R_3C^+, Sn^{2+}, Pb^{2+}	Cu^+, Ag^+, Au^+, Cd^{2+}, Hg^{2+}, Pt^{2+}, Pt^{4+}, Pd^{2+}, CH_3Hg, Tl^+
Bases	NH_3, RNH_2, N_2H_4, H_2O, OH^-, O^{2-}, ROH, RO^-, CO_3^{2-}, SO_4^{2-}, ClO_4^-, F^-, Cl^-, aspartate, glutamate, tyrosine, histidine*	$PhNH_2$, N_3^-, N_2, Br^-	H^-, R^-, C_2H_4, C_6H_6, CN^-, CO, SCN^-, R_3P, R_2S, RSH, RS^-, S^{2-}, I^-, cysteine, methionine

* hard/borderline.

■ To what kind of metal would you expect cysteine to coordinate?

□ Soft ligands tend to coordinate to soft metal ions, including second- and third-row transition-metal ions, particularly those in low oxidation states.

Aspartate, on the other hand, which coordinates through oxygen, is a hard ligand, and will coordinate effectively to hard metal ions such as those of the lighter alkaline earths and the higher oxidation states of transition-metal ions.

2.1.6 Hard/soft ligands in practice

This classification of hardness can be useful when it comes to predicting if a polypeptide will coordinate to a particular metal. The hardness of the coordinating amino acid side chains will determine, to some extent, which metal is preferentially coordinated by that amino acid. Bone is a good example of this. Bone, as we have already seen, contains calcium ions.

■ Is Ca^{2+} classified as a hard or soft acid?

□ Hard acid. Hard acids include the lighter alkali metals and alkaline earth metals, and the first-row transition metals in high oxidation states. Hard acids and bases have low polarisability, whereas the soft acids and bases tend to be more polarisable.

Table 2.3 Chemical content of bone.

Material	Mass/%
fibrous protein (collagen)	≈ 30
hydroxyapatite	≈ 55
$CaCO_3/SiO_2/MgCO_3$ mixture	15

Bone is a complex organic–inorganic composite with a mixture of organic and inorganic components; its chemical make-up is shown in Table 2.3. The organic component, collagen, is a protein made up of three polypeptide chains, with the higher-order structure shown in Figure 2.2. Carboxylate-containing amino acid side chains are regularly spaced along the collagen fibres.

■ Do carboxylate-containing side chains act as hard or soft ligands?

□ Carboxylate coordinates through oxygen; we expect these side chains to be hard, and therefore to coordinate to hard metal ions (Table 2.1).

So, as Ca^{2+} is a hard metal ion, carboxylate groups should, and do, act as good ligands. Indeed, using a combination of a hard ligand with a hard metal ion, and a rigid, regularly spaced array of ligands, collagen is able to **chelate** (coordinate with) calcium ions selectively in the bone-growing process. The whole process is known as **biomineralisation** and is a good example of how amino acid side chains attached to a rigid protein structure can lead to extraordinary metal coordination chemistry: we shall study biomineralisation in more depth in Chapter 6.

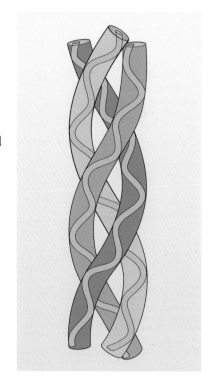

Figure 2.2 The higher-order structure of collagen, showing three intertwined helical polypeptides.

Another feature that is influenced by the hardness of ligands is the stability of a metal ion's oxidation state.

■   Which oxidation state of copper will be favoured by soft and hard ligands?

□   From Tables 2.1 and 2.2, we might expect Cu(I) to have a preference for soft ligands such as the cysteinyl side chains (soft), whereas Cu(II) is a harder metal and is likely to be stabilised by harder ligands such as histidyl side chains (hard/borderline).

■   If a copper ion were coordinated by a mixture of hard and soft amino acid side chains (e.g. two histidyl (hard/borderline) and two cysteinyl (soft) side chains), which would be the more stable copper oxidation state, if either?

□   There is no simple answer to this question because there is clearly a balance between the relative stability of the copper oxidation states, I and II. A copper ion coordinated by two cysteinyl (soft) and two histidyl side chains (hard/borderline) will be expected to show little preference for either of the two oxidation states.

In biochemical systems this does occur. There are classes of copper-containing proteins in which the copper ion in the protein is coordinated by a mixture of hard and soft ligands, such that neither oxidation state is stabilised over the other. These proteins can both accept and donate an electron; in other words, the copper oxidation state can change. We will discuss these proteins in further depth in Chapter 7.

You should note that this hard/soft acid–base theory should only be used as an aid to *predict the preference* of a specific ligand with particular metals. A 'hard' acid, for example, may also form a complex with a 'soft' base, as in the example of Fe^{3+}, which will form a complex with cysteine.

2.1.7   Selectivity in metal binding: stereochemistry and stability

The concept that ligands in a fixed arrangement chelate a metal selectively is an important one in biochemistry. In individual proteins, amino acid side chains are held together in a regular, rigid arrangement by the higher-order structure of the protein chain to which they are attached. We have already met this concept briefly in the example of bone in the previous section. Another good example of the importance of rigid structure in proteins comes from an enzyme found in human blood plasma called **copper–zinc superoxide dismutase (SOD)**. The structure of the protein's active site is shown in Figure 2.3a. It contains one Zn^{2+} ion and one Cu^{2+} ion. The Zn^{2+} ion is coordinated by two neutral histidyl side chains, one deprotonated histidyl side chain and one aspartyl side chain. The copper(II) ion is coordinated by three neutral histidyl side chains and one deprotonated histidyl side chain. (The histidyl side chain is based on the heterocyclic compound imidazole, **2.6**. Hence, the deprotonated side chain ring is known as an imidazolate, **2.7**.)

2.6

2.7

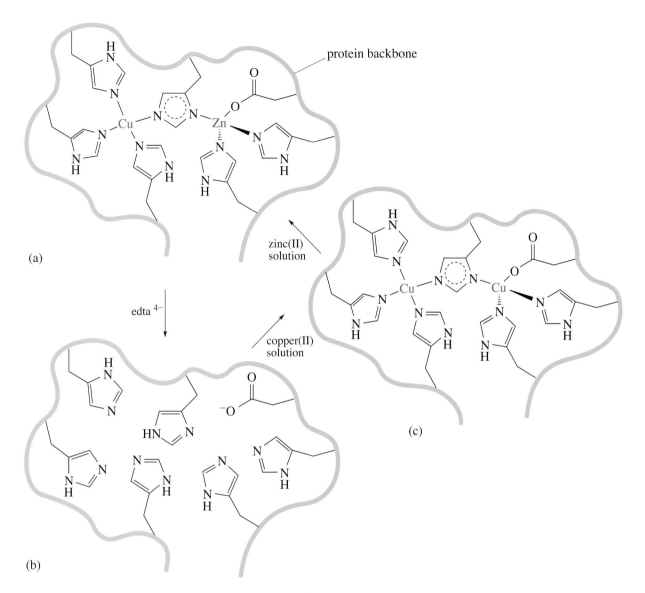

Figure 2.3 (a) Active site of copper–zinc superoxide dismutase showing copper and zinc sites linked with a bridging histidyl ligand; (b) treatment by edta solution to give active site of metal-free protein, showing intact amino acid structure; (c) active site after treatment of metal-free protein with aqueous copper(II) solution. (Note that the coordination geometry of the copper site is distorted square planar.)

As shown in Figure 2.3b, the metal ions can be removed from the protein by treating it with edta, a strong chelating agent, at pH 7; this gives an intact, metal-free (apo) protein.

■ What happens to the deprotonated bridging histidyl under these conditions and why?

☐ In the absence of metal ions, the original deprotonated histidyl side chain is reprotonated, as expected from the pK_a value of 14 of neutral histidine.

This demonstrates the earlier point that deprotonation can be induced by metal ions that coordinate to the deprotonated amino acid side chain, even at a pH lower than the pK_a of the conjugate acid.

Subsequent treatment of the metal-free protein with an aqueous Cu^{2+} solution gives a protein with Cu^{2+} ions in both of the original Cu^{2+} and Zn^{2+} sites (Figure 2.4c). Finally, treatment of the copper-only protein with an aqueous Zn^{2+} solution replaces the copper ion in the zinc(II) site with a Zn^{2+} ion; this gives back the original copper–zinc protein (Figure 2.3a).

■ From the **Irving–Williams series**, given below, which shows the relative order of the **stability constant** for a metal complex (independent of ligand) across the first transition series (and beyond), would you predict that copper(II) would be replaced by zinc(II)?

$$Ba^{2+} < Sr^{2+} < Ca^{2+} < Mg^{2+} < Mn^{2+} < Fe^{2+} < Co^{2+} < Ni^{2+} < Cu^{2+} > Zn^{2+}$$

You will find when we are talking about the oxidation state of a transition metal species in general terms, we will represent the oxidation state in brackets as Roman numerals, as we have done here.

□ No. In fact, what is remarkable about the replacement of copper(II) by zinc(II) is that it is contrary to the stability expected from the Irving–Williams series. The Irving–Williams series predicts that copper(II) complexes are more stable than the corresponding zinc(II) complexes.

Clearly, for some reason, one of the sites must have a strong preference for a Zn^{2+} ion.

■ What do you think might be the reason for this reversal of trends, such that Zn^{2+} does replace Cu^{2+} in the protein?

□ The answer lies in the nature of the amino acid side chains of the ligands and their particular spatial arrangement at the zinc(II) site. Both the 'size' of the site and the geometry of the ligands may play a role.

The spatial arrangement of the two neutral histidyl side chains, aspartyl side chain and the deprotonated histidyl side chain is such that only a particular metal ion with the correct size will give a very stable complex. Zn^{2+} (with an ionic radius of 74 pm) may be a better fit than Cu^{2+} (ionic radius 71 pm), although the difference in ionic radii is not large.

Another factor in the stability of a metal ion complex is the coordination geometry of the metal ion. Most zinc(II) complexes exhibit tetrahedral coordination geometries, whereas most copper(II) complexes exhibit either square-planar, square-pyramidal or distorted octahedral coordination geometries. From these observations, it is reasonable to assume that the most stable zinc(II) complexes have tetrahedral geometries, and that a spatial arrangement of amino acid side chains that gives a tetrahedral coordination geometry at the metal ion will favour the coordination of Zn^{2+} over Cu^{2+} ions. Indeed, the zinc(II) site in the protein has a near-tetrahedral coordination geometry around the metal ion. It is most likely therefore for SOD that stereochemical considerations are the driving force for the site preference. We will return to consider the biochemical properties of SOD in Chapter 8.

These examples serve to give some indication of how proteins can fine-tune the coordination environment in order to bind preferentially the metals they need for biological activity. We shall return to metal chelation and selection in Chapter 3, where we shall examine in detail how iron ions (as Fe^{2+} or Fe^{3+}) are chelated selectively within biochemical systems.

2.2 Macrocyclic ligands

The next important class of ligands in living systems is the family of **macro-cyclic ligands**, the most significant group of which are the **tetrapyrroles**. These have the basic skeleton shown in Figure 2.4a for **porphyrin**. A wide range of tetrapyrroles is known, which are distinguished from each other by the number of double bonds contained within the unit, and the types of side chain attached to the central unit; see Figure 2.4b–d for some more examples.

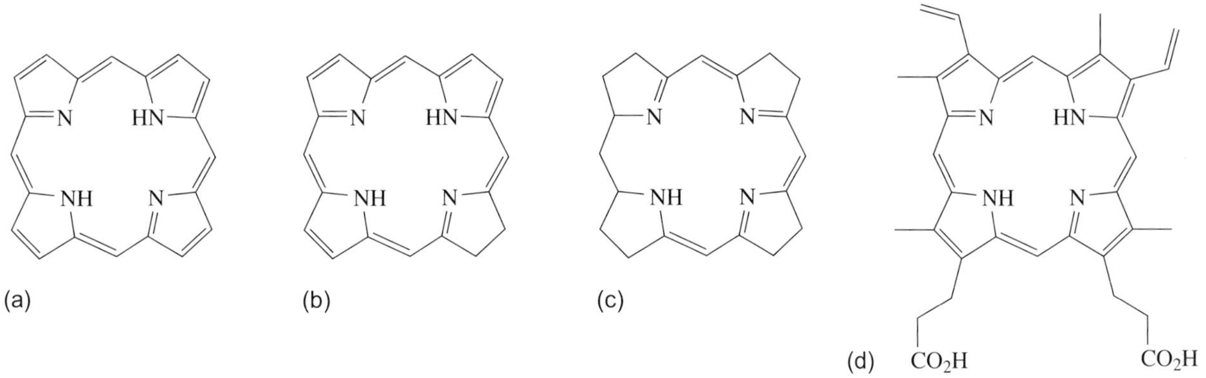

Figure 2.4 Various tetrapyrrole molecules. (a) Porphyrin group, showing the basic tetrapyrrole structure, (b) chlorin, (c) corrin, (d) protoporphyrin IX.

Figure 2.5 shows a porphyrin ligand chelating an iron ion; note that two of the ligand's protons have been displaced in the complexation by iron(II) and that the porphyrin ligand has a formal charge of −2. This is another example of a metal-ion induced deprotonation, as seen with certain amino acid side chains. This haem group, as we shall see in Chapter 8, is an integral part of haemoglobin, the oxygen-transport protein in blood.

- ■ Porphyrins are **polydentate** ligands, that is more than one atom in the ligand binds to the metal ion. What is the **denticity** (the number of atoms in a single ligand that bind to the central metal atom) of the doubly deprotonated iron–porphyrin group shown in Figure 2.5?

- ☐ Four. Porphyrin (as indeed are the other tetrapyrroles) is a tetradentate ligand. (This can be denoted by κ^4, where κ (kappa) describes the denticity of a ligand.)

The tetrapyrroles provide a planar (or very nearly planar) structure and are exceptionally stable. This stability is due in part to the **macrocyclic effect** where an *n*-dentate macrocycle forms a more stable complex than the equivalent open chain *n*-dentate ligand.

- ■ Would you expect the tetrapyrroles to be selective ligands for metal coordination?

- ☐ Yes, the size of the 'hole' in macrocyclic ligands will contribute to their selectivity for metal coordination. This is especially so for the tetrapyrroles because of their rigidity due to the double bonds.

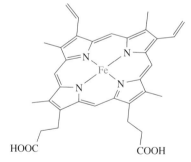

Figure 2.5 Iron–porphyrin group (also known as a haem group).

(Structural data and theoretical calculations suggest that metal ions of radius 60–70 pm are ideal for coordination.)

The same phenomenon in a chemical context is exemplified with the synthetic crown ligands (see Figure 2.6). For example, in 12-crown-4, the crown size is so small that it chelates the small Li^+ ion very effectively. Other ions, like Na^+ and K^+, although chemically very similar, are too big to be 'encapsulated' by the crown ligand; compared to the Li^+–12-crown-4 complex, the analogous Na^+ and K^+ complexes are of low stability.

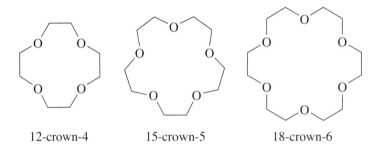

12-crown-4   15-crown-5   18-crown-6

Figure 2.6   Structures of crown ligands; 12-crown-4 (selective for Li^+), 15-crown-5 (selective for Na^+), 18-crown-6 (selective for K^+). The name 12-crown-4, for example, shows that there are 12 atoms in the ring and four oxygen atoms.

You will come across many examples of the tetrapyrroles and other macrocyclic ligands throughout this book.

2.3   Nucleic acids as ligands

Another important class of ligands in bioinorganic chemistry is the **nucleic acids**. Figure 2.7 depicts a **nucleotide**, which is simply a **nucleobase** connected to a ribose (sugar)/triphosphate unit. Nucleotides are the building blocks of DNA (deoxyribonucleic acid) and RNA (ribonucleic acid), i.e. they are polymers of nucleotide monomers. There are several atoms within a nucleotide that in principle could coordinate to a metal ion.

■   Through which atom will the phosphate groups coordinate? Will they be hard or soft, and which metals do you think they will coordinate?

☐   Phosphate coordinates through oxygen and so is classified as hard. The hard phosphate groups form stable complexes with hard metal ions like Ca^{2+} and Mg^{2+}. In fact, single-crystal X-ray diffraction studies suggest that Mg^{2+} coordination by phosphate – in which the phosphate groups are on the outside of the helix – stabilises (along with extensive hydrogen-bonding) the overall DNA double-helix structure.

The potentially coordinating nitrogen atoms of nucleobases are softer than the phosphate oxygens. Although not confirmed, it is believed that the nitrogen atoms of the nucleotides are able to coordinate many of the softer metals, such as Cd^{2+} and Hg^{2+}. These nitrogen atoms are an essential feature in maintaining the DNA double-helix structure: any coordination to metals disrupts the helical structure of DNA, and potentially leads to DNA destruction and diseases such as cancer.

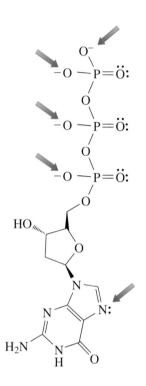

Figure 2.7   Structure of guanosine triphosphate. The arrows indicate potential coordinating atoms on the molecule. Only one nitrogen atom of the nucleobase is a sufficiently strong Lewis base to act as a potential coordinating atom.

Don't forget that there are questions on the companion website which you can use to test your understanding of the material covered in this chapter.

3   Metal uptake – increasing the solubility of iron

3   Metal uptake – increasing the solubility of iron

We considered the issue of relative abundance of the inorganic elements in Chapter 1. Despite the apparently high abundance of some minerals, several challenges exist that can limit the acquisition of metal ions by living organisms. Obtaining sufficient metal nutrients can be a problem. As many as two billion people worldwide suffer from malnutrition owing to deficiencies of trace nutrients such as iron and zinc, for example. An understanding of metal uptake systems is important for the successful treatment of the health disorders related to metal deficiency, such as anaemia, or indeed to metal overload, as in the example of patients suffering from haemochromatosis. This chapter will consider the acquisition of metals, illustrated using the example of iron.

3.1   Introduction

So, what are the major obstacles limiting the acquisition of metal ions? Well, one of the first things to consider is the bioavailability of metals as we discussed in Chapter 1. How much of a particular element is available will depend not just on the abundance of that element on Earth but also where it is found and in what form. Efficient 'capture' of inorganic elements is essential if an organism is to survive. As much of the chemistry of life occurs in aqueous media, the **solubility** of metals (as metal salts) is an important factor to consider. When they dissolve, these metal salts dissociate into aqueous ions. Solubility will depend on the nature of the metal salt and also the pH and temperature of the solution. Human life is restricted to a narrow range of conditions, generally a pH of 7 and a temperature of 303 K. As we shall see in this chapter, the chemical properties of metal ions such as their solubility, reactivity and redox chemistry, together with their ability to form metal ion complexes (and in turn the stability of these complexes), will greatly affect the availability of metal ions to organisms.

The ability of organisms to capture inorganic ions selectively is exemplified by the concentration/dilution factors of the metal ions in human blood plasma, the source of metal ions in the body, shown in Table 3.1. It can be seen that the concentrations of ions in human blood plasma are markedly different from the concentration of the same ion in seawater. (The basis for correlating seawater concentrations with human plasma concentrations stems from the assumption that all living species derive from the primordial organisms that once lived in the oceans. The concentration of inorganic ions in human plasma is, therefore (and rather crudely), assumed to be the same as the concentration of inorganic ions within these primordial organisms.) These differences in concentration are an indirect measure of the active uptake of inorganic ions by humans, and organisms in general.

Table 3.1 Comparison of approximate concentrations of metal ions in seawater with human blood plasma; 'concentration/dilution factor' is [human plasma]/[seawater], and gives an indication as to how the particular element is 'concentrated' or 'diluted' by humans.

Element	[Seawater] /mol dm^{-3} × 10^{-8}	[Human plasma] /mol dm^{-3} × 10^{-8}	Concentration/dilution factor
sodium	4.6×10^7	2×10^5	~4×10^{-3}
magnesium	5.3×10^6	9×10^4	~0.02
potassium	9.7×10^5	2×10^5	~0.2
calcium	1.0×10^6	1×10^6	~1
vanadium	4.0	17.7	~4
chromium	0.4	5.0	~14
manganese	0.7	10.9	~15
iron	0.005–2*	2230	1100–450 000
cobalt	0.7	0.0025	3.6×10^{-3}
nickel	0.5	10.4	~21
copper	1.0	1650	1650
zinc	8.0	1720	215
molybdenum	10.0	1000	1000

* Dependent on pH, which may vary locally from its average value of 8.2, and on the anion content of seawater.

For some elements, such as iron and copper, the difference in concentrations is very high, with significantly higher concentrations of the metal ions in plasma. Certainly, in humans at least, these ions are actively taken up in an energy-consuming (endothermic) process. Metal ions are transported into the body's cells resulting in the enrichment seen. On the other hand, for some metal ions a dilution effect is seen. For example, cobalt ions have a relatively high concentration in seawater but a very low concentration in plasma. So, how exactly are metal ions absorbed by the body and why might their concentrations differ for different ions?

The biochemical processes that are involved in metal uptake are complicated. There are four key steps involved:

- trace quantities of the metal ions must first be captured by an organism from its local environment
- the metal cation should be absorbed by the organism in a selective manner
- the cation needs to cross the cell membrane into a cell
- the metal ions need to be transported to where they are required.

We will consider the first two steps in this chapter, with transport of the metal ions being covered in Chapter 4. Although the actual mechanisms for metal uptake and subsequent transport are not entirely clear, the best understood example is that of iron. Indeed, the uptake of iron provides perhaps one of the most extreme examples of the challenges that organisms must overcome to obtain metal ions and will be used here to illustrate some of the general principles of metal uptake.

3.2 Principles of iron chemistry: the problems of iron uptake

Iron has a high natural abundance. It is the second most abundant metallic element by mass in the Earth's crust (4.1%). This high abundance might suggest that iron is readily available to organisms, but let's consider its chemistry.

■ What are the main oxidation states of iron?

☐ Naturally occurring iron exists primarily in two oxidation states, +2 and +3, but in the presence of O_2 the most stable oxidation state is +3.

We saw previously (Table 3.1) that the iron concentration in seawater is very low, roughly 5×10^{-11} to 2×10^{-8} mol dm^{-3}.

■ If an aqueous pale-green coloured solution of iron(II) nitrate, $Fe(NO_3)_2$, at pH 7 is exposed to air, a brown-coloured precipitate soon forms. On standing, the precipitate, which contains iron(III), settles to the bottom of the vessel. What do you think has happened?

☐ The precipitate is mostly hydrated iron(III) oxide, $Fe_2O_3.nH_2O$ (rust).

■ So, how is the hydrated iron(III) oxide formed?

☐ In aqueous solution, iron(II) will react with O_2 to give iron(III):

$$4Fe^{2+}(aq) + O_2(g) + 4H^+(aq) = 4Fe^{3+}(aq) + 2H_2O(l)$$

(3.1)

Iron(III) is a strong Lewis acid, and will react with water in the following series of **hydrolysis** reactions to give a mixture of highly insoluble $Fe(OH)_3$ and $Fe_2O_3.3H_2O$, which appear as a brown-coloured precipitate:

$$Fe^{3+}(aq) + H_2O(l) = [Fe(OH)]^{2+}(aq) + H^+(aq)$$

(3.2)

$$[Fe(OH)]^{2+}(aq) + H_2O(l) = [Fe(OH)_2]^+(aq) + H^+(aq)$$

(3.3)

$$[Fe(OH)_2]^+(aq) + H_2O(l) = Fe(OH)_3(s) + H^+(aq)$$

(3.4)

$$2Fe(OH)_3(s) = Fe_2O_3.3H_2O(s)$$

(3.5)

Notice that a proton is produced in each of the reactions, hence these hydrolysis reactions *must* be dependent on pH (pH = $-\log[H^+]$). We will consider this in the next section.

3.2.1   pH and solubility

At low pH (high H^+), for example say less than pH 1.5, iron(III) is soluble in water, but at pH 7 a precipitate of $Fe(OH)_3$ is formed. $Fe(OH)_3$ is profoundly insoluble, with a solubility product, $K_{sp} = 2 \times 10^{-39}$ mol^4 dm^{-12} (Box 3.1).

Box 3.1   Solubility product

The solubility product is a measure of the solubility of a compound in water at pH 7 (and usually 298 K). For a sparingly soluble salt, M_aX_b, in contact with its saturated aqueous solution,

$$M_aX_b(s) = aM^{b+}(aq) + bX^{a-}(aq)$$

the solubility product is given by:

$$K_{sp} = [M^{b+}(aq)]^a [X^{a-}(aq)]^b$$

The value of the solubility product can also be expressed logarithmically:

$$pK_{sp} = -\log K_{sp}$$

Accordingly, a high value of pK_{sp} or a low value of K_{sp} indicates that the compound is only sparingly soluble. A lower value of pK_{sp} or a higher value of K_{sp} will indicate a higher solubility and hence higher availability to an organism.

■ What is the concentration of $Fe^{3+}(aq)$, $[Fe^{3+}]$, in H_2O at pH 7, assuming that $Fe(OH)_3(s)$ is present? *Hint*: You can obtain $[OH^-]$ from the dissociation constant for water, $K_w = [H^+][OH^-] = 10^{-14}$ mol^2 dm^{-6}.

☐ We know that the solubility product, K_{sp}, of $Fe(OH)_3$ is 2×10^{-39} mol^4 dm^{-12}.

$$K_{sp} = [Fe^{3+}][OH^-]^3$$

Therefore,

$$[Fe^{3+}][OH^-]^3 = 2 \times 10^{-39} \ mol^4 \ dm^{-12}$$

At pH 7: $[H^+] = 10^{-7}$ mol dm^{-3} and $[OH^-] = 10^{-7}$ mol dm^{-3}. ($[OH^-] = K_w/[H^+] = 10^{-14}$ mol^2 $dm^{-6}/10^{-7}$ mol dm^{-3})

Therefore,

$$[Fe^{3+}] = \frac{2 \times 10^{-39} \ mol^4 \ dm^{-12}}{(10^{-7})^3 \ mol^3 \ dm^{-9}} = 2 \times 10^{-18} \ mol \ dm^{-3}$$

From the calculation above, we can see that at pH 7, the concentration of $Fe^{3+}(aq)$ is extremely small: 2×10^{-18} mol dm^{-3}. The value of 5×10^{-11} to 2×10^{-8} mol dm^{-3} for $Fe^{3+}(aq)$ in seawater, given in Table 3.1, is much higher than this.

■ Why do you think the concentration of $Fe^{3+}(aq)$ in seawater is higher than the value you have just calculated?

□ This is due to the presence of other ligands in seawater that can form more soluble iron complexes. For example, chloride, bromide, acetate and nitrate are all potential ligands present in seawater.

Nevertheless, for such an important bioinorganic element, the concentration of $Fe^{3+}(aq)$ in neutral aqueous solution is very low indeed.

This very low concentration is a significant problem when it comes to an organism obtaining iron, as iron in its soluble form is in such short supply. If an organism is to survive, it must have some biochemical means of absorbing the trace amounts of surrounding iron from its environment. The presence of free Fe^{2+} must also be avoided (Box 3.2). In addition, the organism must prevent iron(III) oxide precipitation once the iron is inside the organism, as this may lead to cell damage.

Box 3.2 Fe²⁺ and the superoxide radical

Any *free* iron(II) aqueous ions dissolved within an organism (not combined in a complex or compound) are also potentially dangerous.

■ What are the products of the reaction of a single iron(II) ion with a single molecule of O_2?

□ $Fe^{2+} + O_2 = Fe^{3+} + O_2^{\bullet -}$

The reaction of O_2 with Fe^{2+} gives single-electron oxidation of the Fe^{2+} to Fe^{3+} and the generation of a **superoxide radical anion**, $O_2^{\bullet -}$. This radical anion is highly toxic to biological systems and must be avoided wherever possible. Therefore, the organism must have methods for preventing the formation of free iron(II).

3.2.2 Coordination complexes of iron

The last property of iron we shall examine is the thermodynamic stability of its coordination complexes.

■ Would you classify iron(III) as a hard or a soft metal?

□ Iron(III) is a first-row transition metal in a high oxidation state, and so is classified as a hard metal. As such, it will tend to form stable complexes with hard ligands. Typically, hard ligands contain oxygen and nitrogen as the coordinating atoms.

The hexadentate ethylenediaminetetraacetate ion (edta^{4-}, Figure 3.1a) ligand coordinates through both O and N, and forms an exceptionally stable complex with iron(III) (shown in Figure 3.1b), with a stability constant of about 10^{25} mol^{-1} dm^3. The analogous stability constant with iron(II) is only about 10^{14} mol^{-1} dm^3. Edta^{4-} is an example of a chelating ligand. Such complexes containing chelate rings are usually more thermodynamically stable than similar complexes without rings; this is known as the **chelate effect** (Box 3.3).

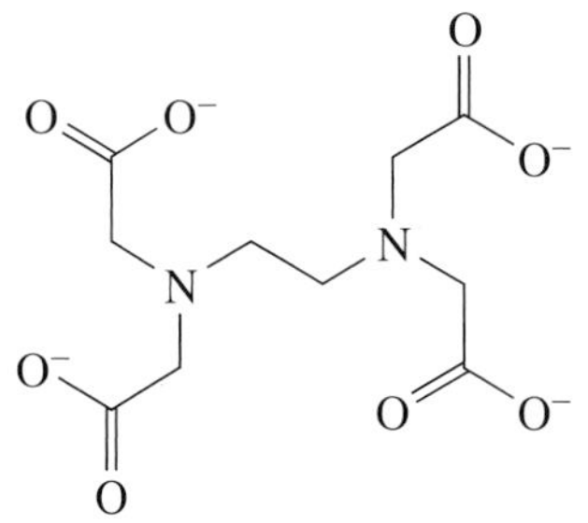

(a) 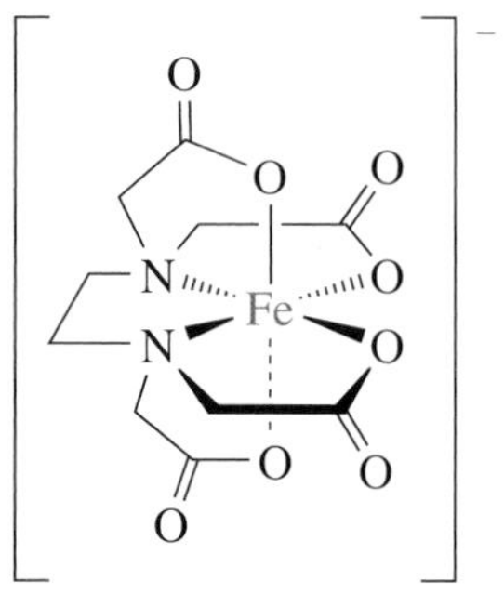   (b)

Figure 3.1 (a) The edta^{4-} anion and (b) its coordination mode to iron. The hexadentate ligand assumes an octahedral arrangement around the metal ion, forming five-membered chelate rings.

Box 3.3   The chelate effect

The chelate effect is observed for pairs of complexes when the coordinating atoms of the monodentate ligands are the same as those of the chelating polydentate ligands, and there is no steric strain in the chelate ring. It is thought that several factors are involved overall in the chelate effect, but that the most influential is the entropy change in the formation reaction. This can be seen in the experimental data given below in Table 3.2 for the formation of two four-coordinate cadmium complexes:

$$Cd^{2+}(aq) + nL = [Cd(L)_n]^{2+}$$

(3.6)

$[Cd(CH_3NH_2)_4]^{2+}$, **3.1**, is coordinated through nitrogen to four monodentate methylamine ligands, whereas $[Cd(en)_2]^{2+}$, **3.2**, is coordinated through nitrogen to two bidentate ethylenediamine ligands. We see that the chelated complex is far more stable than the complex with no rings – by a factor of over 10^4 in this case.

From Table 3.2 it is evident that the enthalpy change for the formation reaction of each complex is very similar, which is to be expected because the atoms involved in each bond, and therefore type of bonding, are similar in each case.

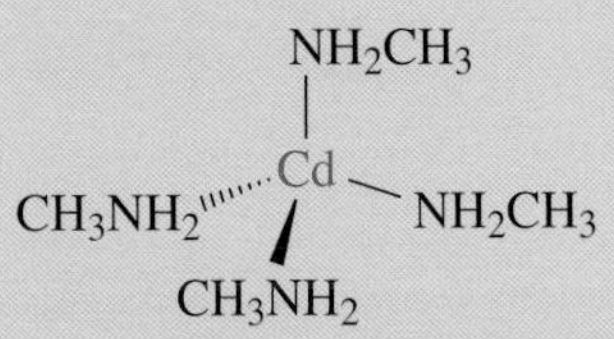

NH$_2$CH$_3$

CH$_3$NH$_2$''''''Cd—NH$_2$CH$_3$

CH$_3$NH$_2$

3.1

NH$_2$

H$_2$N''''''Cd—NH$_2$

H$_2$N

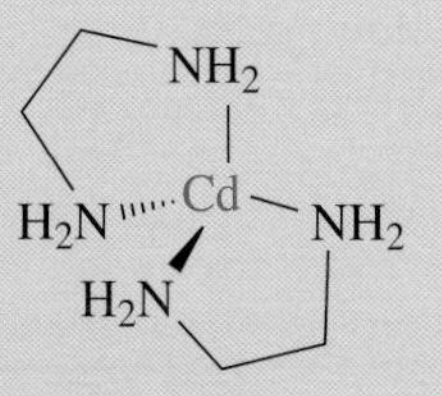

3.2

Table 3.2 Stability constants (given as log K) and thermodynamic data at 298.15 K for some cadmium(II) complexes.

Complex	Log K	$\dfrac{\Delta H_m^\ominus}{kJ\ mol^{-1}}$	$\dfrac{\Delta G_m^\ominus}{kJ\ mol^{-1}}$	$\dfrac{\Delta S_m^\ominus}{J\ K^{-1}\ mol^{-1}}$	$\dfrac{-T\Delta S_m^\ominus}{kJ\ mol^{-1}}$
$[Cd(CH_3NH_2)_4]^{2+}$	6.55	−57.32	−37.41	−66.8	19.91
$[Cd(en)_2]^{2+}$	10.62	−56.48	−60.67	14.1	−4.19

■ What do you notice about the values for the change in entropy?

□ They are very different. The entropy change for the chelated complex is positive, whereas that for the complex with no chelate rings is negative.

The large difference in entropy changes, $\Delta S_m^\ominus$, leads to a big difference in the Gibbs free energy changes for the reactions.

■ How does this lead to a difference in the stability constants of the two complexes?

□ From the relationship $\Delta G_m^\ominus = \Delta H_m^\ominus - T\Delta S_m^\ominus$ we see that a *positive* entropy change will lead to a lower or more negative value for the Gibbs free energy change, $\Delta G_m^\ominus$ of the reaction. $\Delta G_m^\ominus = -2.303RT \log K$, so the more negative is $\Delta G_m^\ominus$, then the larger is K and the more stable is the complex.

The leading question now is, why does this happen? The answer, hinted to above, seems to be that the dominating factor is the entropy of the ligands. Consider what happens when a hexa-aquo metal ion, M, reacts with a complexing monodentate ligand, L:

$$[M(H_2O)_6]^{n+} + 6L = [ML_6]^{n+} + 6H_2O \tag{3.7}$$

there is no change in the number of molecules in solution: the six water molecules are displaced by six ligand molecules.

■ What is the significant difference when the same reaction takes place with bidentate ligands?

□ In a similar reaction with bidentate ligands, L–L,

$$[M(H_2O)_6]^{n+} + 3L\text{–}L = [M(L\text{–}L)_3]^{n+} + 6H_2O$$

(3.8)

there is a net increase of three molecules in solution.

Now you should be familiar with the fact that for gas–solid reactions, a greater increase in the number of gas molecules in a reaction leads to a greater entropy change. Something similar happens here, but in solution. The increase in the number of independent molecules in the chelate reaction leads to a more positive $\Delta S_m^{\ominus}$.

The chelate effect is usually at a maximum for five- and six-membered rings: smaller rings tend to suffer from strain, and in larger rings the two coordinating atoms act as though they are independent of each other (i.e. there is no energy advantage compared to two monodentate ligands).

As we saw in Chapter 2, macrocyclic ligands, such as porphyrin rings, tend to show similar stabilising effects to polydentate ligands. In the particular case of macrocyclic ligands, this is also known as the macrocyclic effect.

3.2.3   A final note

Thus, the chemical properties of iron demand that organisms have efficient and effective methods of obtaining and then controlling iron. Several different pathways exist for the uptake of metals, and, indeed, may even be present in the same organism, such is the need for iron. It is worth noting that there is a constant competition for the available iron in the natural world. In many cases, the availability of iron is the determining factor in whether an organism can proliferate or not. For example, it is believed that the low concentration of iron in seawater limits the amount of plankton that can grow. Specifically, organisms must be able to (i) solubilise and assimilate iron in their local environment, and (ii) protect the iron once it has been absorbed. As we shall see in the next few sections, the most common methods that an organism uses to mobilise the iron, or indeed other metals, in a more soluble form are chelation, reduction and/or changing the pH.

3.3   Iron uptake by bacteria

Nearly all organisms are able to take up iron. However, only a handful of organisms have had their iron-uptake chemistry studied in any detail. The organism that has received most attention (other than human, which will be considered in Chapter 4) is a single-cell, **prokaryotic** bacterium (found in the human large intestine and elsewhere) called *Escherichia coli* (abbreviated to *E. coli*).

There are many harmless strains of the *E. coli* bacterium; the ones found naturally in the human gut are useful because they synthesise several vitamins of the B-complex and vitamin K. However, there are also over 100 pathogenic (i.e. disease-causing) strains of the bacteria. The most infamous is probably *E. coli* O157:H7, which is very virulent. This strain can find its way into the human food chain (from the intestines of cattle where it is thought to originate), and it causes severe food poisoning due to the toxins excreted by the bacteria. The toxins are absorbed from the gut into the bloodstream; damage to the kidneys occurs, which may eventually result in death, particularly for the very old or young.

The reason that this bacterium has been so thoroughly studied is that it is relatively easy to grow and study colonies of it in the laboratory. The iron-uptake mechanism in *E. coli* is known in a fair degree of detail.

E. coli obtains its iron in a remarkable fashion. Each *E. coli* bacterium within a colony secretes small molecules that are capable of specifically chelating iron. These small molecules are known as **siderophores** (from the Greek for iron carriers; pronounced 'sid-air-o-fores'). Several types of siderophore are known, each capable of chelating iron in a stable iron–siderophore complex. The structures of two such siderophores are shown in Figure 3.2.

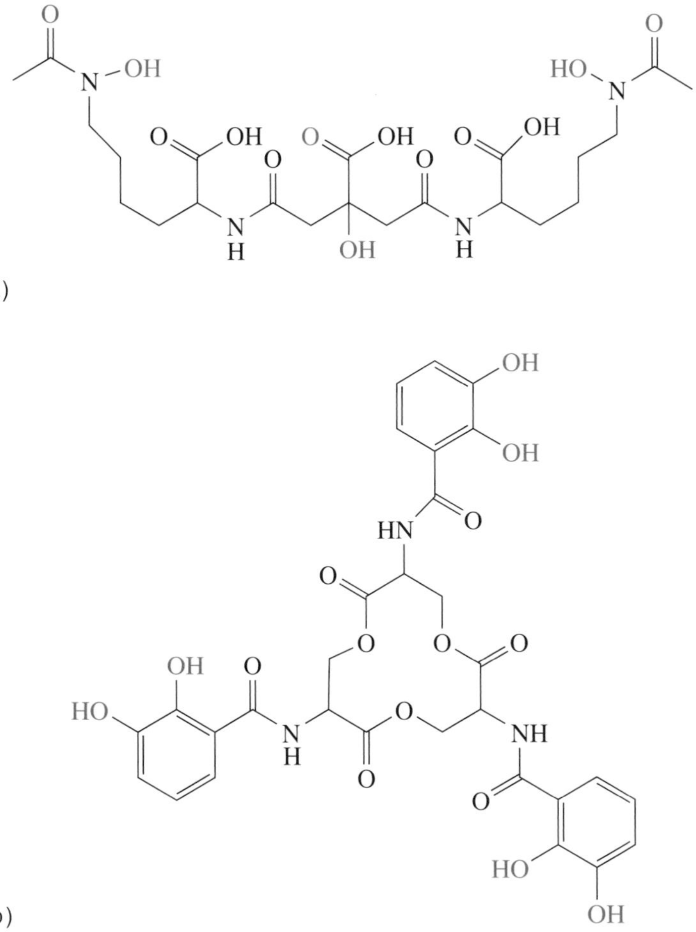

(a)

(b)

Figure 3.2 Structure of the siderophores: (a) aerobactin and (b) enterobactin; coordinating atoms and acidic hydrogens are printed in red.

3.3.1 How siderophores aid uptake

We shall examine the properties of one of the siderophores in more detail below. For all the siderophores, however, *modus operandi* is to be secreted from the bacterium, to chelate an iron(III) ion selectively in a stable complex, and then to be re-absorbed by a bacterium (not necessarily the original bacterium) as the iron(III)–siderophore complex (see Figure 3.3 for a schematic representation).

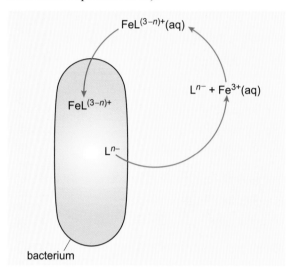

Figure 3.3 Schematic diagram of iron uptake by a siderophore, L.

■ Derive an expression for the concentration of an iron(III)–siderophore complex in terms of its stability constant. Explain why an iron(III)–siderophore complex needs to have a very high stability constant to be biochemically useful.

☐ By writing the equation for the formation of the complex, we can then derive an expression for the concentration of an iron(III)–siderophore complex in terms of its stability constant:

$$Fe^{3+}(aq) + sid^{n-}(aq) = [Fe\text{–}sid^{(3-\,n)+}(aq)] \tag{3.9}$$

(You can use the abbreviation 'sid' to represent the siderophore.)

$$K_{+3} = \frac{[[Fe\text{–}sid^{(3-n)+}(aq)]]}{[Fe^{3+}(aq)][(sid^{n-}(aq)]} \tag{3.10}$$

where K_{+3} is the stability constant for the iron(III) complex. Rearranging Equation 3.10 gives:

$$[[Fe\text{–}sid^{(3-n)+}(aq)]] = K_{+3}[Fe^{3+}(aq)][sid^{n-}(aq)] \tag{3.11}$$

There are two reasons why a high stability constant is required for efficient transport of iron to a bacterium:

First, the value of $[[\text{Fe(III)}\text{–}\text{sid}^{(3-n)+}(\text{aq})]]$ must be significant; this is because the bacterium will have a much better statistical chance of absorbing the iron(III)–siderophore complex if it is in relatively high concentration.

Second, knowing that iron(III) is in short supply and its concentration in water at pH 7 is very low, any organism that can competitively chelate the available iron will have a better chance of survival.

We have shown that the greater the stability constant, the higher will be the concentration of the [iron(III)–siderophore] complex, thus providing the optimum conditions for the transport of the iron. Let's take a look at the rough values of $[\text{Fe}^{3+}(\text{aq})]$ and $[\text{sid}^{n-}(\text{aq})]$ available in the environment. We know that $[\text{Fe}^{3+}(\text{aq})]$ cannot be high and may be as low as 10^{-18} mol dm^{-3} due to the insolubility of many iron(III) compounds. Also, the value of $[\text{sid}^{n-}(\text{aq})]$ cannot be high (that is, probably much less than 10^{-12} mol dm^{-3}), as the bacterium can only ever synthesise a small amount of the siderophore. What this all means is that for the value of $[[\text{Fe}\text{–}\text{sid}^{(3-n)+}(\text{aq})]]$ to be significant, the value of the stability constant, K_{+3}, must be *extremely* high. In other words, for this system to be feasible, the equilibrium for the complex formation must lie very heavily to the right.

3.3.2 Selectivity of siderophores

Another requirement of the siderophore ligand is that it must be selective for iron(III). This means that the stability constant for the [iron(III)–siderophore] complex must be much greater than the stability constants of the siderophore complexes with other metals (including iron(II)).

Why is selectivity so important? Firstly, if the siderophore were not selective for iron(III), high concentrations of other metal ions (M^{m+}) would easily displace the iron(III) from any iron(III)–siderophore complex, according to the equation

$$M^{m+}(\text{aq}) + [\text{Fe}\text{–}\text{sid}^{(3-n)+}(\text{aq})] = [\text{M}\text{–}\text{sid}^{(m-n)+}(\text{aq})] + \text{Fe}^{3+}(\text{aq}) \qquad (3.12)$$

If this equilibrium lay to the right, then iron could not be obtained in any great amounts by the bacterium. Secondly, the bacterium's biochemical systems for absorbing iron should not be a route for the absorption of toxic metal ions, such as mercury and cadmium.

So, how does a siderophore achieve this high degree of iron(III) selectivity? The question can be partly answered by examining the structures of the siderophores in Figure 3.2. We can see that the siderophores are analogous to simpler organic molecules which can coordinate directly to iron(III) ion in a similar fashion, as indicated in Reactions 3.13 and 3.14:

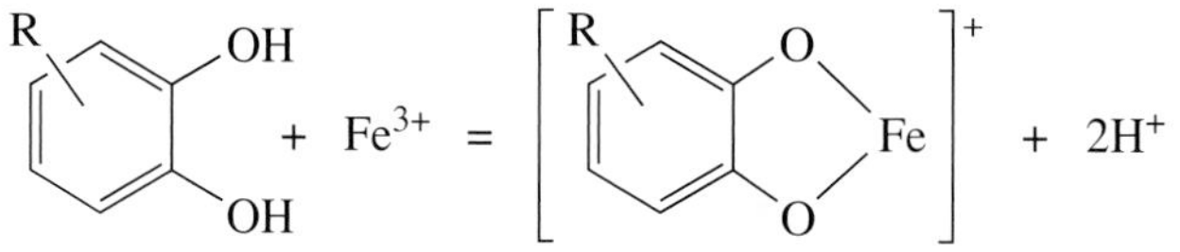

3.3

(3.13)

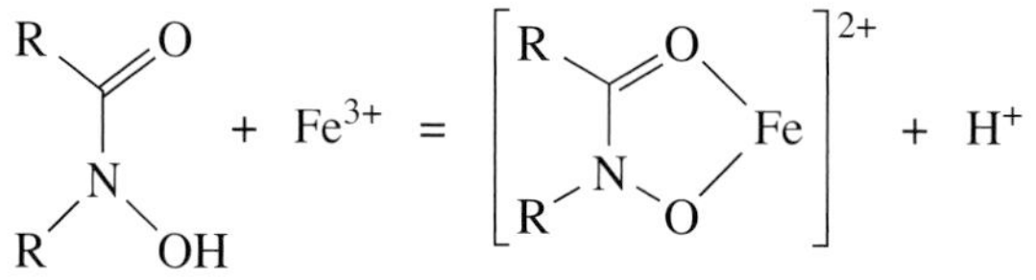

3.4

(3.14)

- ■ Are the coordinating groups of the ligands in Reactions 3.13 and 3.14 hard or soft?

- ☐ All the groups shown are hard ligands and, as such, form stable complexes with hard metals, like iron(III) and aluminium(III).

These groups give their names to types of siderophore. For example, enterobactin is a **catecholate siderophore** (Figure 3.2b) and aerobactin is a **hydroxamate siderophore** (Figure 3.2a).

The groups in structures **3.3** and **3.4**, 1,2-dihydroxybenzene (trivial name **catechol**, pronounced 'kat-a-kol') and hydroxamic acids, respectively, form particularly stable complexes with iron(III), because they are chelating groups. The chelate 'bite' of these two groups is just about right to form a very stable complex with iron(III).

3.3.3 Enterobactin

Of all the siderophores, the one that has received the most attention is **enterobactin**, shown in Figure 3.2b. The reason for this is that enterobactin forms an exceedingly stable complex with iron(III); in fact, it is the most stable, soluble iron(III) complex that is known. The stability constant of fully deprotonated enterobactin with iron(III) is extremely high at about 10^{49} mol^{-1} dm^3. (*Note*: the definition of stability constant assumes an equilibrium reaction in aqueous solution between a hydrated metal ion and ligand(s), such that we would write enterobactin in its ionised, i.e. fully deprotonated, form.)

What are the chemical and structural features of enterobactin that give it such a high stability constant with iron(III)? To answer this question we need to examine the structure of enterobactin in more detail. Figure 3.4a shows the structure of the iron(III)–enterobactin complex.

- ■ Which type of group coordinates to the iron in the complex?

- ☐ The iron atom is chelated by three, deprotonated catechol (known as catecholate) groups.

Notice that iron(III) complexation displaces the six protons on the catechol oxygen atoms of enterobactin, so, overall, enterobactin is a hexadentate ligand

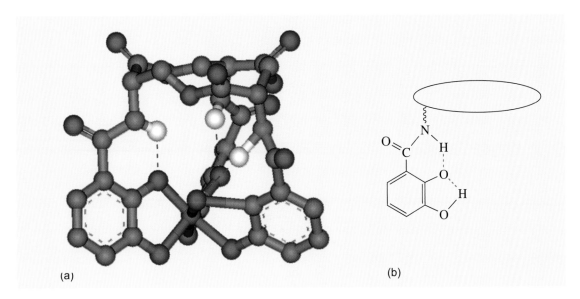

Figure 3.4 Three-dimensional structure of (a) iron–enterobactin complex (the hydrogen atoms other than those involved in H bonding have been omitted for clarity) and (b) structure of one of the partially dehydrogenated catechol groups connected to the serine ring via an amide linkage.

providing six coordinating atoms to the iron. The catechol rings are attached to each other via a 12-membered ring of carbon and oxygen atoms. This ring is a serine trimer, condensed together as follows:

$$3 \left(\begin{array}{c} CH_2OH \\ | \\ H_2N-C\cdots CO_2H \\ | \\ H \end{array} \right) \rightleftharpoons \text{(serine ester ring)} + 3H_2O \qquad (3.15)$$

Rather than forming the normal peptide C(O)–NH bond between the individual amino acid molecules, the serines are linked via ester bonds, whereby the ester is formed between the –OH of one serine side chain and the –CO$_2$H group of another serine:

$$\text{(serine)} + \text{(serine)} \qquad (3.16)$$

serine ester dimer

The result is a 12-membered ring (known as a triserine ring), with three NH_2 groups pointing away from the ring. These NH_2 groups are all pointing to the *same* side of the imaginary plane formed by the triserine ring (this is because natural serine exists as a single enantiomer). To complete the enterobactin structure, three catechol groups are attached to the NH_2 groups of the triserine ring via amide, C(O)–NH, linkages. The overall three-dimensional structure of enterobactin shows a triserine ring, to which three catechol groups are attached via the nitrogen atoms, all linked to the same side of the ring. In partially deprotonated enterobactin the conformation of the molecule is such that the oxygen atoms of the catechol groups turn into the centre, as shown schematically in Figure 3.4b for one of the catechol groups.

■   What type of interaction is likely between the NH of a serine group and the oxygen atom of a catechol ring?

☐   There is a hydrogen bond between the NH of the each serine group and the oxygen atom of the attached catechol ring (shown as dotted lines in Figure 3.4).

These interactions impose further structural rigidity by preventing each catechol ring from rotating freely, so that not only are the catechol groups all linked to the same side of the triserine ring, but all the catechol oxygen atoms face towards the centre of the ligand.

We see that the partially deprotonated enterobactin ligand is actually rather rigid in its structure, with all six coordinating oxygen atoms of the catechol groups held in position to bind an iron(III) ion (that is, the catechol groups have the same relative positions both *before* and *after* the binding of iron(III)). This rigid arrangement of functional groups before the metal has been chelated is known as **ligand preorganisation**; in other words, the three-dimensional structure of the ligand hardly changes on complex formation. What this means in practice is that a metal ion of a particular size (and charge) forms a particularly stable complex with the preorganised ligand. Iron(III) has the correct size and charge to form a very stable complex with the preorganised enterobactin ligand, and, therefore, is chelated selectively by enterobactin. This is exactly the same phenomenon as the selective chelation of alkali metal ions by different sizes of crown ethers that we met previously in Chapter 2. A small crown ligand, like 12-crown-4 (Reaction 3.17), will form stable complexes with a small metal ion, Li^+, whereas a larger crown, such as 18-crown-6, will form stable complexes with a larger metal ion, K^+ (Reaction 3.18):

12-crown-4

$$+ \ Li^+ \longrightarrow \qquad\qquad (3.17)$$

(3.18)

3.3.4 Removal of iron from enterobactin

We will consider how the iron(III)–enterobactin complex is transported into a cell in Chapter 4. However, before temporarily leaving siderophores and iron uptake by bacteria, it is worth asking the question: how does *E. coli* remove the iron from such a stable complex as the iron(III)–enterobactin once it has been absorbed into the cell? At some point the metal must be released for use by other proteins within the cell.

One possible mechanism for the removal of iron from the complex may be found if we look back to Reaction 3.15. The rigid, three-dimensional structure of the triserine ring of enterobactin is the main reason why enterobactin is such an effective ligand. If the structure of the ring is destroyed, enterobactin loses much of its chelating power. In fact, this is what is believed to happen to the iron(III)–enterobactin complex once it has been absorbed by *E. coli*. Enzymes called **esterases** (so-named because they catalyse the decomposition and formation of esters) are thought to hydrolyse the triserine ring of enterobactin in the reverse reaction of Reaction 3.15. This breaks up the triserine ring structure, and the stability constant of the resulting $[\text{Fe(catechol)}_3]$ complex is much lower than 10^{49} mol^{-1} dm^3. Iron is then relatively easily removed from this complex, so that it can be used in the bacterium cell. It has been found, however, that 'model' ligands with related structures to enterobactin, but lacking the ester bonds, were still effective in promoting iron uptake and release, suggesting that a second release mechanism may (also) be important. It has been proposed that this involves reduction of Fe(III) in the complex to Fe(II).

■ The stability constant of a siderophore with iron(II) is in the order of 10^8 mol^{-1} dm^3. How would the reduction of Fe(III) to Fe(II) facilitate release of the iron.

☐ The siderophore binds weakly to Fe(II) and so on reduction, the iron will be readily displaced from the siderophore for use in the bacterial cell.

Although reduction of the Fe(III)–enterobactin complex is not favoured at neutral pH, protonation, for example at lower pH, is believed to make reduction easier. A third possibility is that there is a specific iron-binding protein present within the cytosol of the cell, which strips the iron from the iron(III)–enterobactin complex and prevents the complex from being transported back out of the cell.

3.4   Iron uptake by plants

The example above highlighted one way in which organisms can acquire iron, namely by using chelation of the iron. We will now look at some of the other ways that the solubility, and hence acquisition, of iron (and indeed other metals) can be improved, illustrated with the example of iron uptake in plants. Plants are, of course, the main source of iron in our own diet, either directly or via the diet of animals whose meat we eat.

The inorganic chemical elements known to be essential for plant growth (macronutrients and micronutrients) are listed in Tables 3.3 and 3.4 and include a group of transition metals, notably iron, manganese, zinc, copper and molybdenum, for example. Plants only need tiny concentrations of these elements (usually less than 10^{-4} g per gram of dry matter). Deficiencies of these nutrients may arise for a variety of reasons – high rainfall leading to extensive leaching of light soils, for example, or overly intensive agriculture – but we shall concentrate here on the effect of soil pH on the availability of iron.

Table 3.3 Examples of inorganic macronutrients for plants, showing the form in which they are absorbed, the average relative concentration in dry tissues and some examples of their important functions. Forms absorbed less commonly are shown in brackets.

Element	Form absorbed by plants	Mean relative concentration in dry tissue	Important functions (not necessarily exhaustive)
nitrogen (N)	NO_3^- nitrate (NH_4^+ ammonium)	1000	component of proteins and nucleotides
potassium (K)	K^+	250	osmoregulation (maintains internal balance between water and dissolved minerals); maintenance of electrochemical equilibria; regulation of enzyme activity: e.g. in protein synthesis, ATPases; effects on protein conformation
calcium (Ca)	Ca^{2+}	125	stabilises cell walls and membranes; messenger with important roles in cellular control and coordination
magnesium (Mg)	Mg^{2+}	80	constituent of enzymes, e.g. ATPases and chlorophyll; role in regulation of pH and charge balance within cells
phosphorus (P)	$H_2PO_4^-$ (HPO_4^{2-}) PO_4^{3-} phosphates	60	constituent of nucleic acids, phospholipids, ATP and ADP; role in protein synthesis, membranes and energy transfer as well as regulatory role through phosphorylation reactions
sulfur (S)	SO_4^{2-} sulfate	30	component of proteins and glutathione (an important antioxidant)

Table 3.4 Examples of micronutrients for plants, showing the form in which they are absorbed, the average relative concentration in dry tissues and some examples of their important functions. Forms absorbed less commonly are shown in brackets. Note that Na, Si and Ni are not essential for all plants.

Element	Form absorbed by plants	Mean relative concentration in dry tissue	Important functions (not necessarily exhaustive)
chlorine (Cl)	Cl^- chloride	3	osmoregulation, electrochemical equilibria and enzyme regulation, e.g. in photosynthesis
iron (Fe)	Fe^{2+} (Fe^{3+})	2	component of redox systems: electron carriers, enzymes, e.g. cytochromes, catalases; also needed for protein synthesis
boron (B)	H_3BO_3 boric acid or BO_3^{3-} borate	2	precise function uncertain, but seems to play role in cell elongation, nucleic acid synthesis and membrane function
manganese (Mn)	Mn^{2+}	1	role in photosynthesis; component of the antioxidant superoxide dismutase
zinc (Zn)	Zn^{2+}	0.3	component of enzymes, e.g. carbonic anhydrase
copper (Cu)	Cu^+ (Cu^{2+})	0.1	e.g. component of plastocyanin (an electron transfer protein); role in electron transfer
molybdenum (Mo)	MoO_4^{2-} molybdate	1×10^{-3}	component of nitrate reductase and so has a key role in nitrogen metabolism
sodium (Na)	Na^+	variable	essential in a few, but not all, plants and requirements never as high as in animals
silicon (Si)	SiO_3^{2-} silicate	variable	essential in some (e.g. grasses, as cell wall component), but not all, plants
nickel (Ni)	Ni^{2+}	variable	essential in some, but not all, plants (e.g. present in enzymes in beans)

3.4.1 Iron deficiency and pH

For reasons that are not fully understood, iron deficiency impairs the production of chlorophyll, amongst other things. This is revealed by a yellowing of plant leaves called **chlorosis**, a condition that is particularly marked in sensitive species of the rose family – certain fruit trees, for example – and in some ornamental plants, such as rhododendrons and other *Ericaceae*, when grown on soils of high pH. We will now consider the underlying reason for this problem and the most effective ways of alleviating it.

In Section 3.1 we met the requirement for bioavailability of iron in a soluble form. The mineral nutrients for plants are generally obtained from solution in the soil surrounding the plants' roots. Ions are absorbed, along with water, from the dilute solution that surrounds soil particles. Figure 3.5 shows how the availability of certain nutrients changes with the pH of the soil.

■ How does the availability of iron vary with pH?

□ Iron appears to be increasingly available with decreasing pH (increasingly acid solution), which agrees with the discussion in Section 3.2.

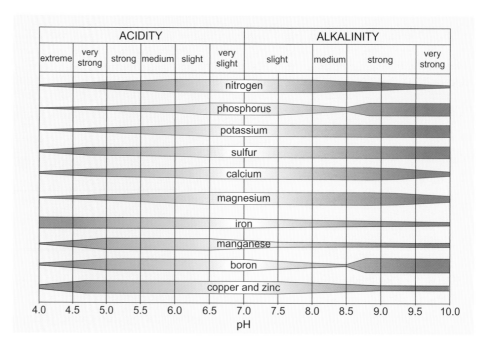

	ACIDITY					ALKALINITY				
extreme	very strong	strong	medium	slight	very slight	slight	medium	strong		very strong

nitrogen
phosphorus
potassium
sulfur
calcium
magnesium
iron
manganese
boron
copper and zinc

| 4.0 | 4.5 | 5.0 | 5.5 | 6.0 | 6.5 | 7.0 | 7.5 | 8.0 | 8.5 | 9.0 | 9.5 | 10.0 |

pH

Figure 3.5 The relationship between the pH of the soil and the availability of mineral nutrients; the wider the bar, the greater the availability.

The pH of soil can affect ion availability in several ways. For example, some ions can precipitate out of solution as insoluble salts which makes them unavailable to plants, e.g. $Fe(OH)_3$. In addition, as in the example of iron, many elements can be present in the soil in different oxidation states, such as Fe^{2+} and Fe^{3+}, but often only one of these is readily taken up by roots. We have met the idea that complex formation or the formation of sparingly soluble compounds affects the relative stability of these ions. Further, the stability of complexes containing different oxidation states may change as the pH changes. We will now quantify these observations.

Imagine that we have a solution of $Fe^{2+}(aq)$ in dilute sulfuric acid. The solution has been prepared not in air, but in an atmosphere of nitrogen and stoppered so that it contains no oxygen. The situation is shown in Figure 3.6a.

Imagine now that the stopper is removed, admitting oxygen. No colour change is observed (Figure 3.6b). Is this consistent with the standard electrode potentials for iron?

$$Fe^{3+}(aq) + e = Fe^{2+}(aq) \quad E^{\ominus} = 0.77 \text{ V} \tag{3.19}$$

$$\tfrac{1}{2}O_2(g) + 2H^+(aq) + 2e = H_2O(l) \quad E^{\ominus} = 1.23 \text{ V} \tag{3.20}$$

No. Thermodynamically we would expect that $Fe^{2+}(aq)$ would be oxidised by oxygen and aqueous hydrogen ions to $Fe^{3+}(aq)$; however, the observation that no oxidation seems to occur in the timescale of the experiment suggests that the reaction is slow.

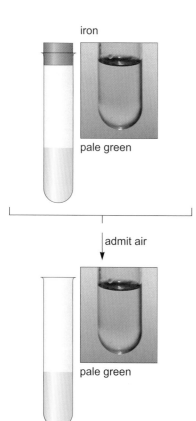

iron

pale green

admit air

pale green

Figure 3.6 Solution of metals in dilute sulfuric acid: (a) $Fe^{2+}(aq)$ prepared by dissolution of iron under nitrogen gas; (b) the effect of admitting air to the solution in (a).

If the acidified solution of $Fe^{2+}(aq)$ is made now alkaline (pH 14), the initial precipitate is pale green, but quickly darkens on standing in air. Is this observation consistent with the information collected in Table 3.5 and the redox potentials below?

$$Fe(OH)_3(s) + e = Fe(OH)_2(s) + OH^-(aq) \quad E^\ominus = -0.58 \text{ V} \tag{3.21}$$

$$\tfrac{1}{2}O_2(g) + 2H^+(aq) + 2e = H_2O(l) \quad E^\ominus = (1.23 - 0.0592pH) \text{ V} \tag{3.22}$$

Table 3.5 Properties of hydroxides of iron.

Ion	Appearance	K_{sp}
$Fe(OH)_2$	Pale green precipitate*	10^{-16}
$Fe(OH)_3$	Brown precipitate	10^{-39}

* Colourless when completely pure, but green as normally precipitated.

Making the solution alkaline leads initially to a precipitate of $Fe(OH)_2$ ($K_{sp} = 10^{-16}$). At pH 14, the potential of the oxygen electrode is $E = (1.23 - 0.0592 \times 14) = 0.40$ V. This is more positive than the corresponding value for the iron couple (Equation 3.21), so aerial oxidation of $Fe(OH)_2$ to $Fe(OH)_3$ is thermodynamically favourable in alkaline solution as, indeed, is the corresponding oxidation of $Fe^{2+}(aq)$ to $Fe^{3+}(aq)$ in acid. In other words, oxygen is thermodynamically capable of oxidising iron(II) to iron(III) at all values of pH. The crucial point is that the process is very much faster in alkali (for test-tube quantities it takes about 5 minutes) than in acid, so here the observed 'stabilisation' of iron(III) with increasing pH rests on kinetic rather than thermodynamic factors.

Thus, we have established that although aerial oxidation of iron(II) to iron(III) is thermodynamically favourable at all pH values, it is faster under neutral or alkaline conditions than in acid. In other words, oxidation of iron(II) to iron(III) is favoured by a high pH, albeit because of kinetic rather than thermodynamic factors.

Now, the pH of soils varies between about 4 and 10. At the lower end of this range, most of the iron will be present as iron(II), whether generated by weathering from rocks or through the bioreduction of iron(III) by microorganisms. The presence of iron(II) ensures a significant concentration of iron in the moisture associated with the soil, because even so-called 'insoluble' compounds of iron(II), such as $Fe(OH)_2$, have solubility products that are not too small (around 10^{-16}). But if a soil has a higher pH (in the range 7–10, say), conversion of iron(II) into iron(III) by aerial oxidation will be relatively fast, especially if the soil has an open, well-aerated texture. This ensures that the iron soon ends up as the much more insoluble hydrated oxide of iron(III), $Fe(OH)_3$ – recall from Section 3.2, that even at a neutral pH of 7, this corresponds to a concentration of $Fe^{3+}(aq)$ of 2×10^{-18} mol dm^{-3}. It is no wonder then, that high pH, combined with good aeration, can result in iron deficiency.

3.4.2   Overcoming iron deficiency

In principle, there are a number of ways that a plant has of tackling this problem. For example, several species that absorb iron effectively even as it becomes deficient, do so largely because their roots are able to release reductants and H^+ ions into the local environment; the more deficient the conditions are, the more efficiently this process occurs.

Iron appears in general to be taken up by roots as Fe^{2+}. A reducing agent (a ferric reductase), together with H^+ ions, is released at the root surface, reducing Fe^{3+} to Fe^{2+}, thereby increasing the concentration of soluble iron.

- ■   How will a decrease in the pH in the local environment around the roots further affect the solubility of iron?

- ☐   Clearly from the calculation above, the maximum concentration of soluble iron will increase at lower pH (that is, higher concentration of H^+ and so lower concentration of OH^-).

This **constitutive mechanism** (where the roots of a plant release a ferric reductase and H^+ ions) is sufficient to deliver adequate iron to most plants in healthy soil (Figure 3.7a); however, in iron-deficient soil (high pH), additional mechanisms exist to aid a plant's uptake of iron.

In some flowering plants, the plant responds to low iron levels by a more efficient release of ferric reductase and H^+ into the soil. In addition, some of these plants also excrete iron-binding ligands and soluble reductants, such as phenols, into the soil. This further increases the solubility and hence uptake of iron (as Fe(II)) by the plant. The absorption process at the root surface involves transport of the iron as Fe(II) into the plant. This releases the ligands for complexing with further Fe^{3+} in the soil particles.

In other plants, including for example grasses and grains, rather than increasing the constitutive system if iron is in short supply, an entirely new mechanism for acquisition of iron is induced. Similar to the siderophores excreted by bacteria, these plants release **phytosiderophores** (or phytometallophores, recognising that they chelate most transition metals) into the soil. These non-protein forming amino acids (an example is mugineic acid, shown in structure **3.5** and chelated with iron in structure **3.6**) are able to chelate iron and indeed other metals, forming stable complexes with Fe^{3+} and other metal cations. These chelates keep the metal available in the soil solution, and hence able to diffuse to the root surface. The Fe(III)–phytosiderophore complex is then taken up into the root cells via specific transporters in the membrane of the cell. We will consider this mechanism and indeed the different routes by which metal ions are transported into cells in general in Chapter 4. Suffice it to say here that this mechanism provides an additional route whereby iron can be acquired by these plants – shown schematically in Figure 3.7b. This explains the remarkable adaptation of these plants, such as the cereals, wheat and oats, to the high-pH soils that are

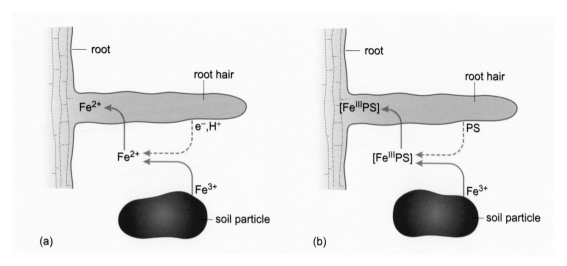

Figure 3.7 Schematic representation of (a) uptake of iron as Fe(II) and (b) enhanced uptake for Fe(III)–phytosiderophores (PS) under conditions of iron deficiency.

usually found in the semi-arid cereal-producing plains in the world. Interestingly the synthesis of the phytosiderophores appears to be quickly suppressed when the plants are restored to adequate iron levels. In addition, although they will chelate most transition metals, the production of phytosiderophores is not induced by a deficiency in other metals, with the exception of zinc, illustrating again the particular demand for iron.

3.5 + Fe^{3+} = 3.6

3.4.3 Artificial enrichment – how gardeners can overcome nutrient deficiency

In addition to the natural mechanisms adopted by plants to increase their iron uptake, there are several methods that gardeners can use to prevent iron deficiency. The first, and simplest, is to add soluble iron compounds, such as $FeSO_4.7H_2O$ to the soil. However, since such compounds are as subject to oxidation and precipitation as is naturally occurring iron(II), this approach is unlikely to be very efficient in practice. A second strategy is to lower the soil pH, thereby both increasing $[Fe^{3+}(aq)]$ and slowing down the oxidation of soluble iron(II). In practice, this is usually achieved by adding sulfur, which is oxidised to sulfate by the action of soil bacteria: the overall effect can be represented (in highly simplified form) as

$$S + \tfrac{3}{2}O_2 + H_2O = 2H^+ + SO_4{}^{2-} \tag{3.23}$$

Although effective (albeit on a fairly short-term basis), this is a rather drastic approach, as changing pH can (and does) affect both the supply of other nutrients, and the activity of soil organisms. Indeed, nutrients – especially micronutrients – can become 'too' available at low pH, reaching concentrations at which they become toxic to plants.

■   Can you suggest another way of increasing the concentration of soluble iron – without changing the pH?

☐   Add to the soil a ligand that forms a stable, soluble complex of iron(III).

As you have seen previously in Section 3.2 (Figure 3.1), a reagent that forms an exceptionally stable complex with $Fe^{3+}(aq)$ is the hexadentate ligand $edta^{4-}$. The stability constant is around 10^{25} for the reaction

$$Fe^{3+}(aq) + edta^{4-}(aq) = [Fe(edta)]^-(aq) \tag{3.24}$$

At pH 8, soil solution has a concentration $[Fe^{3+}(aq)]$ of 10^{-21} mol dm^{-3} (as determined by Equilibrium 3.25)

$$Fe(OH)_3(s) = Fe^{3+}(aq) + 3OH^-(aq) \tag{3.25}$$

Suppose that $edta^{4-}$ is added such that its concentration is subsequently maintained at 10^{-5} mol dm^{-3}.

■   What is $[[Fe(edta)]^-]$ when both Equilibria 3.24 and 3.25 are satisfied?

☐   Equilibrium 3.24 is satisfied by the condition:

$$\frac{[[Fe(edta)]^-]}{[Fe^{3+}][edta^{4-}]} = 10^{25} \text{ mol}^{-1} \text{ dm}^3$$

With $[Fe^{3+}] = 10^{-21}$ mol dm^{-3} (which satisfies Equilibrium 3.25 at pH 8) and $[edta^{4-}] = 10^{-5}$ mol dm^{-3}, $[[Fe(edta)]^-] = 10^{-1}$ mol dm^{-3}.

As suggested by this simple calculation, addition of the chelating ligand $edta^{4-}$ can drastically increase the maximum concentration of soluble iron in the soil solution.

In a very real sense, the chelate structure protects the iron(III) ion from the depredations of its watery environment – and hence effectively prevents the formation of rust. It is interesting to note that nature has adopted much the same strategy to utilise iron at physiological pH in our own bodies. Thus, for example, the oxygen carriers haemoglobin and myoglobin both contain iron encircled by a tetradentate porphyrin ligand: the resulting haem is, in turn, encapsulated within a pocket in the globin protein. You will meet these examples in detail in Chapter 8.

Although solutions containing free edta^{4-}(aq) (or other similar ligands) are sometimes used to rectify iron deficiency, it is more common to apply the Fe(III) complex – usually referred to as **sequestered** iron. Despite their expense, the use of such synthetic chelates is substantial – especially to meet micronutrient deficiencies of citrus and other fruit trees in many parts of the western United States, which have calcareous soils of high pH.

Before leaving this section we should stress that the analysis outlined above presents a grossly oversimplified picture of the soil system. Thus, to suggest that soluble iron(III) in equilibrium with Fe(OH)$_3$ in the soil is in the form of Fe^{3+}(aq) alone gives a misleading impression. Humus in the soil contains chelating ligands, which form complexes with Fe^{3+}(aq) (and other metal cations), thereby increasing the soluble iron. Addition of an artificial chelating agent only supplements the natural chelating powers of the soil – always providing, of course, that steps are taken to maintain the humus content. These natural chelating agents are not well characterised, but there is evidence that phenolic compounds, proteins, amino acids and organic acids all contribute. Further, as you have already seen, siderophores are produced and released into the soil by some bacteria and also a number of fungi. However, their importance to overall plant iron nutrition remains unclear.

3.4.4 Ecological distribution of plants

One final point: plants differ enormously in their ability to tolerate the deficiency of a given micronutrient – or, equally, an excess. Indeed, the ecological distribution of many plants is strongly influenced by the soil pH. Plants adapted for growth on calcareous soils are known as calcicoles (lime-lovers), e.g. cowslip (*Primula veris*) and rock-rose (*Helianthemum nummularium*). Conversely, plants adapted for growth on acid soils are known as calcifuges (lime-haters) and include species typical of upland moors and sandy heaths such as bilberry (*Vaccinium myrtillus*), heathers (e.g. *Calluna vulgaris* and *Erica cinerea*) and rhododendron. Most plants are adapted to grow well at around pH 6.5, at which most essential elements are readily available. If the pH rises above 7.5, as it does in soils overlying chalk and limestone, then elements such as Fe, Mn and Zn become relatively unavailable and so there is a potential for deficiency. At pH values below 6.0, these same micronutrients may reach toxic concentration. Indeed, poorly aerated waterlogged soils of low pH often contain concentrations of Fe^{2+}(aq) that would be toxic to many plants. There is evidence that the roots of plant species tolerant to these conditions can oxidise the iron to the +3 state, and hence render it less available.

Don't forget that there are questions on the companion website which you can use to test your understanding of the material covered in this chapter.

4   Metal transport

In Chapter 3, we considered the strategies that organisms use to acquire metals in a soluble form; this however raises further questions. How does a charged metal ion cross the cell membrane into a cell? How is the metal ion transported to where it is actually needed in an organism? It is these questions that we will address in this chapter. We will start with perhaps one of the most well-known and important examples of metal ion transport, that of the transport of alkali and alkaline-earth metal ions across a cell membrane.

4.1   Transport across a cell membrane – introduction

So far, we have focused primarily on the transition metal elements, particularly iron, and their coordination chemistry. Turning now to the alkali (Group 1) and alkaline-earth (Group 2) elements, we find some of the most abundant ions in biological systems – Na^+, K^+, Ca^{2+} and Mg^{2+} (Box 4.1). These ions are readily soluble and are typically taken up as the aqueous ion. Generally speaking, the complexing ability of alkali and alkaline-earth cations is much weaker and less well studied than that of the transition metals, although they do bind to oxygen atoms in ligands such as carbonyls, carboxylates, phosphate, and sometimes alcohol or water. Octahedral coordination is the most common, although the larger potassium and calcium cations can exhibit coordination numbers up to eight.

Box 4.1   Role of Group 1 and 2 ions in the body

Sodium, potassium, calcium and magnesium are often described as essential minerals – a group of elements we require if we are to stay healthy. In fact they are members of a larger group of about 16 elements that perform vital biochemical functions. Rough guidelines to how much we (assuming we are 'average adults') need of each from our food are so-called Recommended Daily Allowances (RDAs). For the elements of Groups 1 and 2, typical values are: sodium 1600 mg, potassium 3500 mg, calcium 700 mg and magnesium 300 mg. These are relatively high intakes compared to many elements; for example, you only require about 1.2 mg of copper a day, and others such as selenium or molybdenum are only needed at trace (microgram) levels.

But why do we require these elements?

It is important to normal body functioning that sodium and potassium in the blood plasma and the fluid around the cells are tightly regulated; these ions are essential for transmission of electrical impulses along nerves and for muscle contraction. Owing to excessive salt in our diet, we often take in far too much sodium for our needs; at high levels of sodium, the body retains too much water and the volume of body fluids increases, thus raising blood pressure. Interestingly potassium seems to have the opposite effect; the higher the intake, the lower the blood pressure. Studies indicate that higher potassium levels allow the body to deal more effectively with excess sodium. As fruit, such as bananas, and vegetables are good sources of potassium, more fruit and vegetables in our diets are thought to have a beneficial effect on blood pressure.

Calcium ions also play an important role in the transmission of electrical signals along nerves and in the brain, and in muscle contraction; however about 99% of the calcium in our bodies is present as a major constituent of bones and teeth. We mainly obtain the calcium we need from milk and dairy products, but holding on to it is another matter. Various compounds in food can bind to calcium, preventing it from being absorbed from the digestive system into the blood. For example, oxalic acid, which is present in spinach and rhubarb, reacts with calcium to form calcium oxalate. In general it appears that only about 30% of the calcium in food is actually absorbed into the blood; the rest is lost in faeces.

Magnesium is another major constituent of bone and is central to the successful functioning of many biochemical molecules. For example, the functioning of **adenosine triphosphate (ATP)** hinges on the involvement of magnesium.

The next few sections will centre on one example of the biological chemistry of the alkali and alkaline-earth metals; the movement of Na^+, K^+ and Ca^{2+} ions across cell membranes. This forms just part of the incessant traffic of molecules and ions that continually flow into and out of cells, and is essential for life. To a chemist, much of the interest in this area centres on processes which run counter-intuitive to accepted chemical wisdom; namely the movement of ions in aqueous solution against a concentration or voltage gradient and through a hydrophobic environment (in this case, the 'oily' interior of a cell membrane). However, before launching into this topic in detail, Section 4.2 summarises the key features of cells and, in particular, cell membranes of which you will need to be aware.

4.2   Cells and cell membranes – the nuts and bolts

All living things are composed of one or more cells, and the molecular components of all cells are remarkably similar. A generalised animal cell is shown in Figure 4.1a. The integrity is maintained by the cell membrane, inside which lies the **cytoplasm** containing the cell components – **organelles** (each themselves bound by a membrane) surrounded by a fluid substance known as **cytosol**. The most prominent organelle is the **nucleus**. It is estimated that two metres of DNA is tightly packaged within the nucleus of each cell, surrounded by a nuclear membrane. As such the nucleus contains the genetic information necessary for reproduction, growth and metabolism and in many ways is the 'control centre' of the cell. Not all organisms have nuclei however; those that do are said to possess **eukaryotic** cells, and those with their DNA unbound, free within the cell, are termed **prokaryotic** cells. The latter are single-celled organisms such as bacteria.

Other examples of organelles include the oval-shaped **mitochondrion** (often called the 'power house' of the cell), responsible for production of most of the cell's supply of ATP, and the **endoplasmic reticulum (ER)**, a network of membranes formed into tubes and sacs. The ER is mainly responsible for the synthesis of many important biomolecules such as proteins and lipids. It is divided into two types, according to whether it is associated with **ribosomes** (sites of protein synthesis) or not: rough ER is associated with dense globule-shaped ribosomes and is involved in protein synthesis; smooth ER does not contain ribosomes and is involved in lipid synthesis. The proteins synthesised in the rough ER are transported to the **Golgi apparatus**. This structure consists of stacks of flat membrane-like sacs in close proximity to many round 'bags' of membranes of various sizes, termed **vesicles**. Newly synthesised proteins are modified and sorted in the Golgi apparatus through this arrangement of sequential sacs to produce proteins of the correct structure. These fully functional intracellular proteins are ultimately packed into vesicles that are then directed to the appropriate places within the cell. Finally, **lysosomes** are organelles that contain digestive enzymes. They carry out the function of breaking down molecules imported by the cell that are not required, and also degrading waste material or unwanted organelles.

A typical plant cell is shown in Figure 4.1b. It contains similar components to the animal cell; however, in addition, it is surrounded by a **cell wall** (outside the cell membrane), which provides rigidity to the cell. It also contains **chloroplasts**, organelles where photosynthesis occurs, and **vacuoles**, filled with watery solution, which provide structural support for the cell.

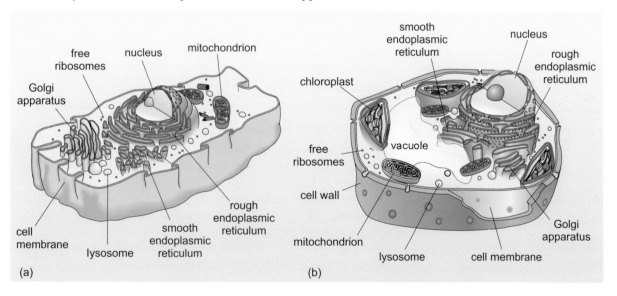

Figure 4.1   A schematic diagram of (a) a typical animal cell and (b) a typical plant cell.

Electron microscopy and X-ray diffraction, coupled with analysis of chemical composition, have revealed that eukaryotic cell membranes are assemblies of lipid and protein molecules. Lipids are composed of polar, **hydrophilic** ('water-loving') heads and long, non-polar, **hydrophobic** ('water-hating') tails. The three main types of lipid in a cell membrane are phospholipids, glycolipids and cholesterol. Phospholipids (generally the most abundant) comprise a phosphate group, which forms the hydrophilic head, and two long hydrocarbon chains, the hydrophobic tails (Figure 4.2a). Glycolipids are similar to phospholipids, but have one or more sugar chains attached (Figure 4.2b); cholesterol also has a polar head – a hydroxyl group, the remainder of the molecule being a hydrocarbon (Figure 4.2c).

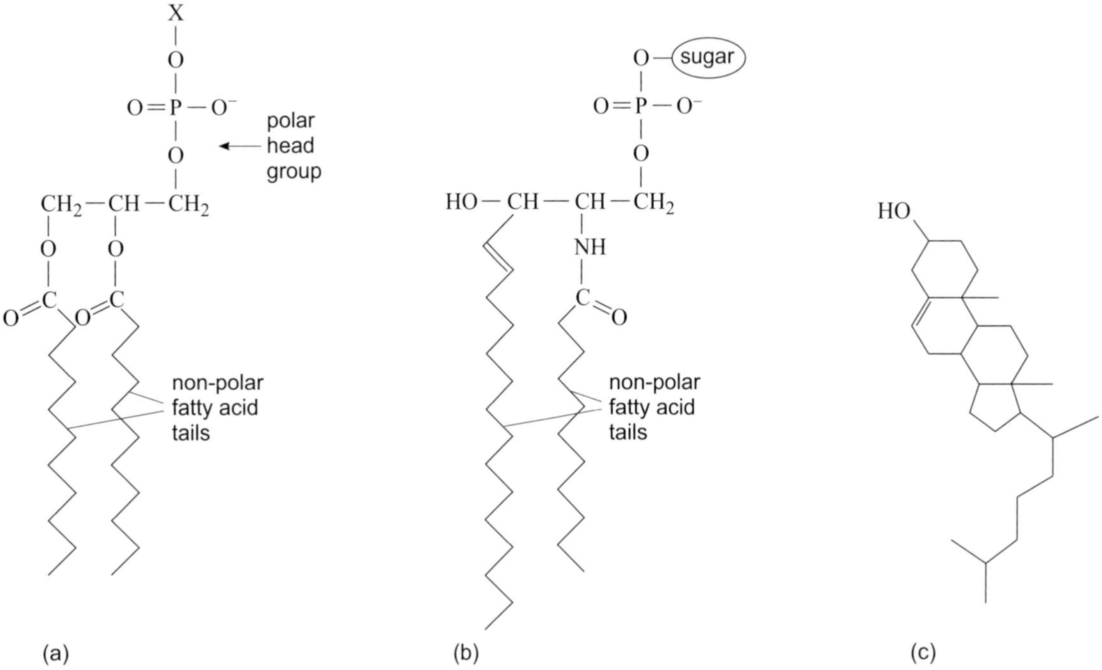

Figure 4.2 (a) Simplified structure of a phospholipid based on glycerol. Note that the substituent 'X' is always polar (and may be positively charged). (b) An example of a glycolipid. (c) The structure of cholesterol.

■ How will phospholipid molecules arrange themselves in an aqueous environment?

□ The phosphate-containing groups will be in contact with water molecules, whereas the hydrophobic chains will form an arrangement whereby they will avoid water.

This behaviour is the driving force for the arrangement of lipid molecules in a cell membrane, which is made up of a double layer (bilayer) of phospholipids. This places the phosphate-containing heads in contact with water on both the inside of the cell (the cytosol) and the outside (the extracellular fluid), whereas the hydrophobic tails are hidden in the interior of the membrane.

Embedded within the lipid bilayer are various proteins. The proportion of protein relative to lipid is different in different membranes. For example, a typical cell membrane is 50% protein (by mass), whereas the internal membranes of mitochondia are around 75% protein. The individual protein molecules are arranged either within the bilayer with parts of the molecule sticking out on either side (transmembrane protein), or attached to either the inside or outside surface. Given that the transmembrane proteins span three regions of a cell (aqueous on the interior and exterior, and non-aqueous within the membrane), structurally these molecules are divided into three domains: two polar sections, one on either side of the membrane, joined by a transmembrane portion which presents non-polar amino acid side chains to the lipid interior.

A schematic representation of the structure of a cell membrane is shown in Figure 4.3. It is important to stress that this is not a rigid structure, but highly dynamic; the protein molecules are often loosely described as being 'dissolved' within the lipid bilayer and are mobile, forming part of a so-called *fluid mosaic.* The fluidity is a consequence of the weak (non-covalent) interactions between the membrane constituents.

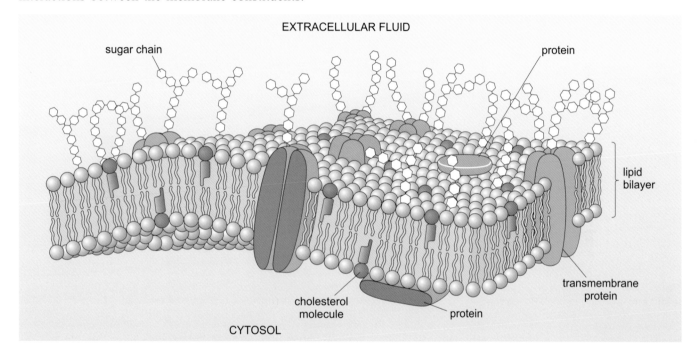

Figure 4.3   The fluid mosaic model of membrane structure.

4.3   Transport across a cell membrane – some key issues

Movement of materials from one side of a cell membrane to another is crucial for controlling the internal environment of a cell. For example, nutrients and (usually) oxygen are required for metabolism and waste products must be able to get out. Consider too the functioning of the nervous system, which requires manipulation of ionic concentrations across the membrane in response to electrical signals.

Cells obtain the molecules or ions they require from the surrounding extracellular fluid, but this is not always straightforward. A lipid bilayer is permeable to small gaseous molecules such as O_2, CO_2 and N_2 and small uncharged polar molecules such as water, urea and ethanol. Such molecules move across the membrane down their concentration gradient (known as passive diffusion, Box 4.2), whereby they dissolve in the phospholipid bilayer, move across it, and then dissolve into the aqueous solution on the other side (it is important to recognise that, unlike the transport mechanisms discussed later in Section 4.4, only lipid molecules are involved in this process). In contrast, the lipid bilayer is much less permeable to large uncharged polar molecules (e.g. glucose or amino acids) or to ions, yet it is vital that species such as these can enter and leave a cell.

Box 4.2 Passive diffusion

The molecules in a liquid or gas are constantly on the move, bouncing around randomly in between collisions with other molecules. They therefore tend to spread out to fill any space that is available.

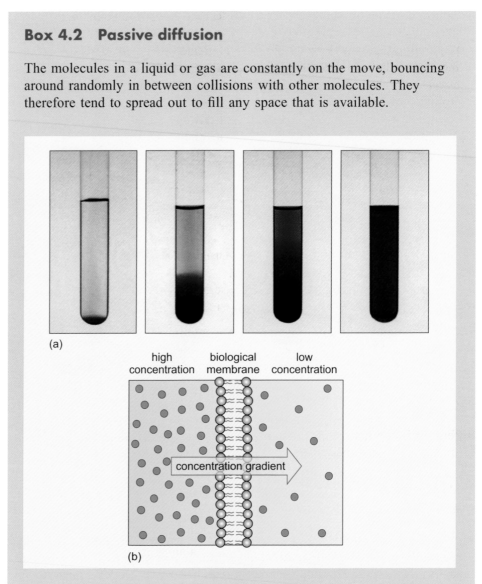

Figure 4.4 (a) Diffusion of purple dye in a test tube containing water: (from left to right) shortly after setting up; after 7 hours; after 24 hours; and after 3 days. (b) Schematic representation of a concentration gradient across a biological membrane, showing the direction in which dissolved molecules diffuse until the concentration equalises on both sides of the membrane.

Diffusion is the movement of molecules from regions of high concentration into regions of lower concentration, until there is an even distribution throughout the available volume.

Think about a drop of dye in a glass of water. Even if you don't stir it, after several days the dye will have spread evenly throughout the water (Figure 4.4a). The same principle applies to movement of molecules across a biological membrane that is *permeable* to those molecules (that is, it cannot act as a barrier against them). Biological membranes are usually permeable to very small molecules such as water, oxygen and carbon dioxide. Movement of these molecules across the membrane occurs by *passive* diffusion (that means no energy expenditure is required) as long as the concentration is greater on one side of the membrane than it is on the other (as in Figure 4.4b). In this situation a *concentration gradient* is said to exist. You could think of molecules 'rolling down' the gradient from the high side to the low side. Unless other forces oppose them, molecules will always diffuse throughout the available space, like the drop of dye, until the concentration gradient is abolished.

4.3.1 Potential gradient

Movement of ions into and out of cells, particularly of Na^+, K^+ and Ca^{2+}, will form the main focus of this section. As these are charged species, in addition to their concentration gradient, a further factor that has a bearing on their travel through the membrane must be considered.

The concentrations of mobile ions inside the cytosol may be very different from those in the extracellular fluid, as you can see in Table 4.1. The Na^+ and K^+ concentrations for a typical mammalian cell differ markedly across the membrane, and even though the Ca^{2+} concentrations appear to be the same, most Ca^{2+} inside the cell is located inside organelles and the concentration in the cytosol is in fact only about 10^{-7} mol dm^{-3}. As you shall see, this difference in the ionic concentrations produces an electrical potential gradient (voltage) across the membrane.

Table 4.1 A comparison of the concentrations of Na^+, K^+ and Ca^{2+} inside and outside a typical mammalian cell.

Cation	Concentration inside cell/ mmol dm^{-3}	Concentration outside cell/ mmol dm^{-3}
Na^+	5–15	145
K^+	140	5
Ca^{2+}	1–2	1–2

Consider the simplified system illustrated in Figure 4.5. KCl(aq) is in two compartments, at different concentrations, separated by a physical membrane. If the membrane between the two compartments is permeable to K^+ but not

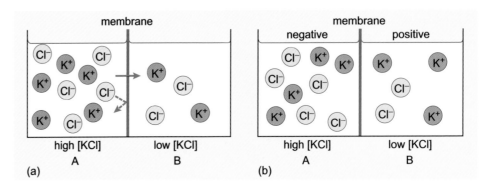

Figure 4.5 Schematic of a system with two compartments separated by a membrane containing KCl(aq) of different concentration (a) initially and (b) after diffusion of K^+ with a potential difference across the membrane.

Cl^-, then the K^+ ions will tend to diffuse down the concentration gradient from compartment A, at higher concentration, to B at lower concentration (Figure 4.5a). As the membrane is not permeable to Cl^-, there will be no movement of Cl^- between the compartments. So, as the positively charged K^+ ions diffuse across the membrane, there will be a net transfer of positive charge from compartment A to B, and B will become electrically positively charged with respect to A. Thus a potential difference or electrical gradient will be set up across the membrane (Figure 4.5b). This process will not continue indefinitely as the electrical gradient will tend to retard further migration of K^+ from A to B. An equilibrium will be established where the driving force from the electrical gradient will be in balance with the driving force due to the concentration gradient. The electrical potential difference at which this occurs is known as the **equilibrium potential** or **Nernst potential**.

The resulting potential difference for a particular ion at equilibrium, V_N, may be calculated using a form of the Nernst equation:

$$V_N = V_{in} - V_{out} = \frac{RT}{zF} \ln\left(\frac{c_{out}}{c_{in}}\right)$$

where V_{out} and V_{in} and c_{out} and c_{in} are the potentials and the ionic concentrations outside (extracellular) and inside (cytosolic) the cell, respectively, z is the ionic charge, F is the Faraday constant (= 96 485 C mol^{-1}), and R is the universal gas constant (= 8.314 J K^{-1} mol^{-1}).

■ Calculate V_N for K^+ and Na^+ (assuming a Na^+ concentration inside the cell of 12 mmol dm^{-3}) at 298 K (*Hint*: 1 V = 1 J C^{-1}).

☐ For Na^+

$$V_N = \left\{\frac{(8.314 \text{ J K}^{-1} \text{ mol}^{-1})(298 \text{ K})}{(1)(96\,485 \text{ C mol}^{-1})}\right\} \ln\left(\frac{145 \text{ mmol dm}^{-3}}{12 \text{ mmol dm}^{-3}}\right) = +64 \text{ mV}$$

For K^+

$$V_{N} = \left\{\frac{(8.314 \text{ J K}^{-1} \text{ mol}^{-1})(298 \text{ K})}{(1)(96\,485 \text{ C mol}^{-1})}\right\} \ln\left(\frac{5 \text{ mmol dm}^{-3}}{140 \text{ mmol dm}^{-3}}\right) = -86 \text{ mV}$$

■ What factor accounts for the large difference between these values of V_N in the calculation?

☐ The very different extracellular and cytosolic concentrations for these two cations; Na^+ has a higher concentration outside the cell, whereas K^+ has a higher concentration inside the cell.

4.3.2 Membrane potential and permeability of ions

In the previous section we have seen that each ion has its own equilibrium potential which depends on the charge and concentration of the ion. However the overall potential difference across the membrane, the **membrane potential**, takes into account the concentrations and charges of all the mobile ions on either side of the membrane.

■ Considering the different ions, what other factor will need to be taken into account when considering their contribution to the overall membrane potential?

☐ The permeability of the membrane to the ions, that is how easily an ion can cross the membrane, will also be an important factor. In the example above, the membrane was only permeable to K^+ ions and not Cl^-. In a cell, the membrane may have very different permeabilities for the different ions, even, for example, between apparently similar ions such as Na^+ and K^+, as we shall now see.

For a cell at rest, a membrane potential of −70 mV is characteristic for many mammalian cell types. Thus the inside of the membrane is negative with respect to the outside and as such will encourage the entry of cations and retard the entry of anions.

Now, let's take a step back and think a little about the *magnitude* of this voltage. At first sight this value might appear rather small, but if you consider that a typical membrane is only about 5 nm thick, then this corresponds to an electric field of 70×10^{-3} V/$(5 \times 10^{-9}$ m$) = 1.4 \times 10^7$ V m^{-1}. A field of this magnitude would be sufficient to cause the electrical breakdown of most solid and liquid materials and indeed is several orders of magnitude higher than the voltage gradients in high voltage electrical power cables.

How then can this extremely thin lipid membrane resist the passage of ions so effectively when the driving forces are so large?

Let us consider the process involved in crossing the membrane.

■ What type of environment does the lipid bilayer of the membrane provide?

☐ The lipid bilayer will be hydrophobic.

The standard enthalpy of hydration is the heat evolved when I mol of gaseous ions becomes surrounded by water molecules (i.e. hydrated) measured under standard conditions.

We start by considering two of the physical properties of the ions themselves. Some values of ionic radii and standard enthalpy of hydration, including those of the ions under consideration, are listed in Table 4.2.

Table 4.2 Ionic radii and enthalpies of hydration of selected ions in water at 25 °C.

Ion	Radius/pm	Hydration enthalpy/10^3 J mol^{-1}
Li$^+$*	60	−520
Na$^+$	96	−405
K$^+$	133	−321
Rb$^+$	148	−300

* Li$^+$, as the smallest ion, produces the largest electric field at its surface; this in turn polarises the local water dipoles to the greatest extent giving the largest hydration energy. Hence on descending the group, the hydration energy decreases.

Cells obtain ions from the surrounding extracellular fluid; these will then need to cross the hydrophobic membrane. This can be likened to stripping the surrounding water ligands from the ion, and placing the bare ion in the lipid. The energy gained by placing it in the lipid is small due to the low electrical polarisability of lipid compared to water. The dominant effect will be the energy required to strip the ion of its water ligands, the reverse enthalpy of hydration of the ion.

■ Which of the two ions, K$^+$ and Na$^+$, would you expect to have the higher permeability?

It is the higher permeability of K$^+$ ions that accounts, at least in part, for the typical value of the membrane potential for a cell at rest at −0.70 mV which is closer to that for K$^+$ (−0.86 mV) than Na$^+$ (+0.64 mV).

□ According to the discussion above, the permeability of an ion will depend on the reverse enthalpy of hydration of the ion. K$^+$, with the lowest enthalpy of hydration, is likely to have a higher permeability than Na$^+$.

In fact, the permeability for both ions is still very low. The probability of these ions surmounting such a large energy barrier to cross the bilayer is negligible, so the thin lipid membrane forms a very effective insulator. In effect the cell membrane is behaving rather like an **electrical capacitor**. This is a device for storing charge, which comprises an insulating sheet sandwiched between two plates of electrically conducting material. In the context of a cell, the hydrophobic interior of the membrane is the insulator whereas the polar head groups together with the surrounding aqueous solution are electrically conducting. This analogy is developed further in Box 4.3, which will tell you more in detail about the origin of the membrane potential or transmembrane voltage.

Box 4.3 A more detailed description of membrane potentials

Consider then, the following simple model of a cell, comprising a 5 μm (5 × 10^{-6} m) radius sphere of univalent salt solution of concentration 0.1 mol dm^{-3}, separated from the surrounding fluid by an insulating cell membrane (Figure 4.6). Relating this to an electrical capacitor, the ability of the membrane to hold the charge is given by a quantity known as **capacitance**, C. The measured capacitance of a typical membrane is about 5×10^{-3} F m^{-2} (the farad, F, is the unit of capacitance).

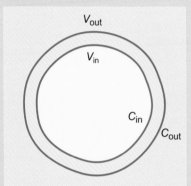

Figure 4.6 A living cell as an electrical capacitor.

The cell has a surface area of $4\pi r^2 = 4 \times 3.14 \times (5 \times 10^{-6})^2$ m$^2 = 3.14 \times 10^{-10}$ m^2. Thus the membrane capacitance for our model is:

$$(3.14 \times 10^{-10} \text{ m}^2) \times (5 \times 10^{-3} \text{ F m}^{-2}) = 1.6 \times 10^{-12} \text{ F}$$

If the charges on the two plates of the capacitor (or in a cell, the two sides of the membrane) are $+Q$ and $-Q$, respectively, and the potential difference between the two is V, then the capacitance, C, is given by $C = Q/V$. Consequently to create a voltage of 70 mV across the membrane requires the transfer of a total charge ($Q = CV$) of $(1.6 \times 10^{-12}$ F$) \times (70 \times 10^{-3}$ V$) = 1.12 \times 10^{-13}$ C across it. (1 C = 1 F V)

■   As a univalent ion carries a charge of 1.602×10^{-19} C, approximately how many ions does this correspond to?

☐   This will correspond to the transfer of 1.12×10^{-13} C/ 1.602×10^{-19} C which is about 7×10^5 univalent ions.

This is actually a very small number when you think that the model cell contains about 3.15×10^{10} cations and the same number of anions (for a concentration of 0.1 mol dm^{-3} in a volume of $4/3\pi r^3 = 5.23 \times 10^{-13}$ dm^3). Thus to create a large transmembrane voltage of 70 mV requires the transfer of *only 1 in every 45 000* cations contained within the model cell across the membrane.

Finally, returning to the concept of a 'gradient', it should now be apparent that when considering the motion of ions across a cell membrane, concentration difference is not the only factor. Overall, the tendency of a particular ion to be transported into, or out of, a cell depends on the *net* effect of its concentration gradient, and the membrane potential; this is known as the **electrochemical gradient** of that ion. The permeability of the membrane to the ion is also important as we shall see further in the next few sections.

4.4 Passive transport across a cell membrane – the role of membrane proteins

So, now you are aware of some of the physical characteristics of cell membranes which impact on ionic transport from one side to the other. How, then, does this process actually take place in a living cell?

You have seen that the permeability of ions through the lipid membrane is low and so passive diffusion is unlikely. The movement of inorganic cations (and anions) across a membrane requires the assistance of proteins, as we shall see shortly, and may proceed either down or against an electrochemical gradient. The former is known as **passive transport**, and the latter is referred to as **active transport** and requires the input of energy. Two types of receptor molecule/protein are involved: channels, which facilitate passive transport only, and carriers, which participate in both passive and active transport.

First let's consider passive transport of ions and how this is mediated by membrane proteins.

4.4.1 Passive transport via ion channels

You need to be careful with semantics here; in contrast to the chemist's definition of a ligand that you have been using routinely so far, in this context, ligand refers to molecules that attach to the protein constituting the channel. They may be viewed as extracellular signalling molecules which trigger events within a cell in response to binding.

An ion channel in a living cell consists of a large protein, or group of proteins, which spans the membrane and is exposed on both sides as illustrated in Figure 4.7. The protein forms a pore in the membrane that allows movement of a substrate across the membrane by diffusion. The channel can be opened or closed by a variety of stimuli (mechanical, electrical or chemical) described by the term 'gated'. Two common examples are **ligand-gated** and **voltage-gated** channels in which the channel opens in response to the binding of a particular ligand or to a change in the local electrical field, respectively. The term 'gated' simply means that the binding of the ligand or the change in voltage opens or closes the channel.

Perhaps one of the most well-known examples of ion channels in the body is their role in signalling, as shown schematically in Figure 4.8, for the sequence that results in muscle contraction. A **neurotransmitter**, acetylcholine, is released from a neuron terminal of a nerve to bind with acetylcholine-gated Na^+ ion channels on the surface of the muscle cell. In normal circumstances this channel is closed; however on binding to the receptor, the pore opens to allow ions to pass across the membrane.

■ What type of gating does this ion channel demonstrate?

□ This is an example of a ligand-gated ion channel.

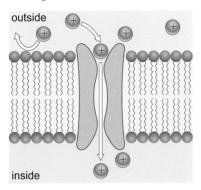

Figure 4.7 Model of membrane transport by ion channels. The ions are shown as red balls.

Passage of the Na^+ ions through the membrane causes an increase in the membrane potential (**depolarisation**) in the region near the receptors (the membrane potential moves from its resting value (−70 mV) close to the equilibrium value for K^+ (−0.86 mV) towards that for Na^+ (+0.64 mV)). This in turn causes voltage-gated Na^+ ion channels in the membrane to open. As more Na^+ ions flow across the membrane, the depolarisation spreads across the membrane causing more Na^+ channels to open. The resulting general depolarisation of the cell leads to the release of Ca^{2+} ions, stored within an

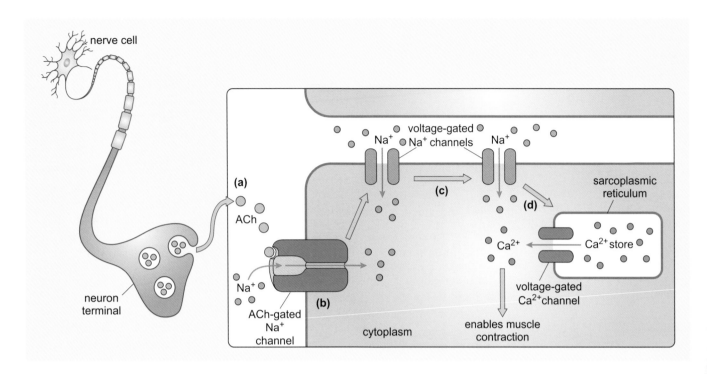

Figure 4.8 Schematic diagram depicting (a) release of acetylcholine (ACh) from a neuron terminal, (b) opening of ACh-gated Na^+ channel, (c) opening of voltage-gated Na^+ ion channels, and (d) release of Ca^{2+} into the muscle cytoplasm via voltage-gated Ca^{2+} channels. The increase in Ca^{2+} concentration in the cytoplasm of the muscle causes the muscle to contract.

organelle in the muscle cell, into the muscle cytoplasm through voltage-gated Ca^{2+} ion channels. Release of Ca^{2+} enables contraction of the muscle. The membrane repolarises with the aid of K^+ ion channels, which open in response to the initial depolarisation but at a slower rate. Unlike the acetylcholine-gated ion channel, which is a non-specific cation channel, these voltage-gated Na^+ and K^+ ion channels are selective to Na^+ and K^+, respectively.

As discussed above, many membrane channels are highly selective, i.e. they can discriminate between cations and anions, and even between different ions with the same charge, for example, potassium ion channels are around 10^4 times more selective for K^+ than for Na^+. This implies that the channel has specific binding sites for K^+ ions. This in turn has some important thermodynamic consequences; to bind to a site, an ion must be (at least partially) dehydrated, and this costs energy. Hence, the stabilisation of the molecular species resulting from binding must be sufficient to compensate for the price paid by dehydration. In addition, it is common, even among highly selective channels, for as many as 10^6–10^7 ions of a given type to pass through a channel every second it is open. Herein lies a puzzle that has taxed researchers in this area; how can these channels be so highly selective and yet exhibit such high throughput rates? Some answers are proposed at the end of this section, but first let's consider the structure of some ion channels.

4.4.2 Structure of ion channels

The complete amino acid sequence of many ion channel proteins has been determined using cloning techniques, and it is clear that there are large families of very similar channels. For example, the amino acid sequences of voltage-gated Na^+-selective channels cloned from different species are very similar so it is reasonable to assume that their three-dimensional structures when spanning the membrane will also be similar. Likewise it has become clear that there is a large family of closely related voltage-gated K^+ channels.

Roderick MacKinnon was awarded a Nobel Prize in Chemistry for this work in 2003. You can listen to his lecture on the Nobel Prize website (MacKinnon, 2003).

In 1998, Roderick MacKinnon and colleagues at Rockefeller University, New York, provided further insight into the structure of ion channels (Doyle et al., 1998). They used X-ray crystallography (Box 4.4) to reveal a detailed picture of the K^+ channel from the bacterium *Streptomyces lividans*. Four subunits were identified and structurally shown to be arranged in the form of an inverted cone (Figure 4.9b). At the wide end is a so-called selectivity filter 1.2 nm in length – a region formed by loops of the polypeptide chain which is so narrow that dehydration of the entering ion is enforced. This is compensated for energetically by coordination of oxygen atoms from amino acid carbonyl groups, which cannot approach the smaller Na^+ ions close enough to provide the necessary stabilisation. When a second ion enters, electrostatic repulsion with the first K^+ ion reduces the time the latter is held and hence movement into the channel takes place. The channel is lined with hydrophobic amino acids and as the ion moves through, it has to overcome an electrostatic energy barrier, which would reach a maximum at the channel centre. At this point however, twin stabilisation mechanisms are thought to operate due to: (i) the existence of an aqueous cavity in the membrane interior; and (ii) the attraction of the negative charges from the carboxyl end of amino acid helices (Figure 4.9c). So the transport mechanism proposed as a result of this study would seem to go a long way towards squaring the high selectivity, high throughput conundrum of ion channels.

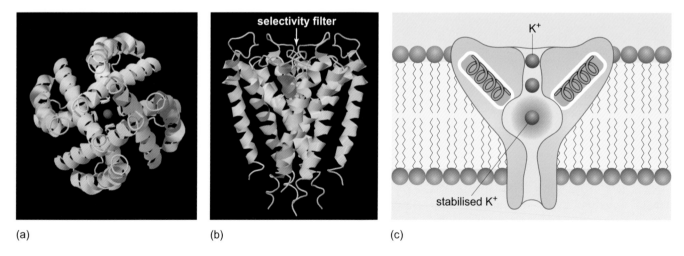

(a) (b) (c)

Figure 4.9 (a) End-on view and (b) side view ribbon representation of the four subunits. (c) A representation of the stabilisation of K^+ by the channel in the middle of the membrane. ((a) and (b) based on pdb file 1bl8 (Doyle et al., 1998).)

Box 4.4 Determining membrane protein structure using X-ray crystallography

The three-dimensional molecular structure determined by X-ray crystallography is known for only a few ion channels as it is difficult to form crystals of membrane proteins. A membrane-spanning protein, such as an ion channel, will only maintain its natural shape if its surroundings in a crystal mimic those it experiences when spanning the cell membrane. For example, if it is exposed to an aqueous fluid, the protein is likely to unfold and take up an entirely different shape from that when spanning the membrane. If the shape of the natural protein were approximated to a circular cylinder, it would require a hydrophobic environment over much of its curved surface to mimic the membrane and a hydrophilic environment at its two ends to mimic the intracellular and extracellular aqueous fluids. Much progress has been made recently in this direction and full structures are now becoming available.

4.4.3 Passive transport by ion carriers

Ion carriers or **ionophores** operate differently from ion channels. A carrier, as its name implies, binds to an ion and releases it on the other side of the membrane, as shown schematically in Figure 4.10.

One of the most well-studied examples is the polypeptide valinomycin; you may recognise this name as that of an antibiotic. Valinomycin is highly selective for potassium ions and acts as an antibiotic, killing bacteria by causing their outer membrane to leak K^+ ions. It has a cyclic structure with 36 atoms in the ring (Figure 4.11a) and chelates K^+ ions, bonding through six of its carbonyl groups to form an octahedral complex (Figure 4.11b). This was demonstrated in a study of the salt $[K(valinomycin)]_2[I_3][I_5]$ by X-ray crystallography, where it was also shown that the ligand conformation in the complex is reinforced by several intramolecular hydrogen bonds.

You may recall that complex formation with macromolecular ligands (such as crown ethers) is a well-established feature of the synthetic coordination chemistry of Group 1 and 2 elements, with numerous complexes having been prepared and characterised.

The high selectivity for K^+ by valinomycin is apparent from the stability constant, K_s, for the potassium–valinomycin complex of 10^6, compared with $K_s = 10^1$ for the sodium–valinomycin complex. This stems from the change in shape of the ligand on coordination; there is less strain on the ligand on bonding to K^+ rather than Na^+. Again, the stabilisation resulting from complex formation compensates for the energy cost of K^+ dehydration.

The uncomplexed ionophores typically possess polar functional groups, aiding their solubility in the aqueous environment outside the cell. On binding a metal ion, the carrier undergoes a **conformational change**, as the functional groups fold inwards, presenting a hydrophobic surface on the exterior that

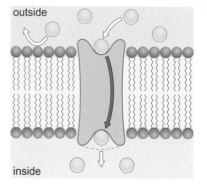

Figure 4.10 Model of membrane transport by carrier proteins. The red arrow denotes movement of the ion binding site to the other side of the membrane. The ions are shown as yellow balls.

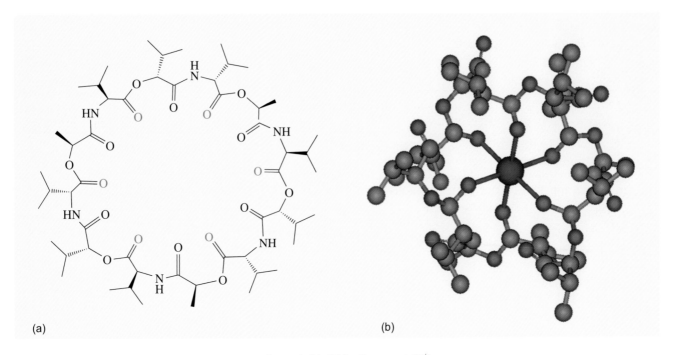

(a)   (b)

Figure 4.11   The structure of (a) valinomycin and (b) [K(valinomycin)]$^+$.

allows it to dissolve in the lipid of the membrane. It can then diffuse across the membrane and deliver that ion to the aqueous phase on the other side. The limiting rate for the transfer is that for the diffusion of the whole complex across the membrane. Indeed, the feature that most clearly distinguishes the operation of an ion carrier from that of an ion channel is its rate of transfer of ions, which, at about 10^3 s^{-1}, is much slower than the rate of 10^6 s^{-1} for the ion channel.

■   What reason can you suggest for the difference in the rate of transfer of ions via carriers and channels?

☐   The slower rate of transfer for carriers is a result of the conformational changes involved on binding that impose considerable kinetic limitations for ion carriers.

Another marked difference is the temperature dependence. Lowering the temperature can markedly slow the carrier, particularly if the change involves a phase change of the lipid into a more ordered form, which inhibits diffusion.

4.4.4   Passive transport and saturation

At low concentrations, transport of ions via either channels or carriers is enhanced by increasing concentrations of an ion, but at high concentrations the transport protein molecules become saturated. At this point the rate of transport does not increase further with increasing ion concentration as seen in Figure 4.12. **Saturation** is characteristic of all protein-mediated transport, making an easy distinction between protein-mediated transport, such as ion channels and carriers, and simple passive diffusion. That the passive movement of ions through, for example, a protein-lined channel has different kinetics from passive diffusion is perhaps surprising. This reflects the fact that the selectivity shown by ion channels is associated with binding of the ion to a region of the protein pore, followed by its dissociation and passage across the membrane.

Both channels and carriers can become saturated, although the saturation of carriers tends to occur at lower concentrations than that of channels (Figure 4.12). These differences in saturation thresholds have been attributed to the differences in the strength of interaction of an ion with the two types of transport proteins; the binding of an ion to a carrier protein is considered to be more tenacious than is the interaction of an ion with a channel protein.

■ In Figure 4.12, what is the significance of the line represented by **A**?

☐ **A** is likely to relate to transport by simple diffusion as there is no evidence that saturation occurs with increasing concentration.

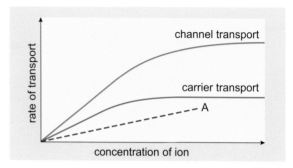

Figure 4.12 Transport kinetics of channels and carriers. Channels allow faster rates of ion movement and saturate at a higher concentration of ion than do carriers.

4.5 Active transport – ion pumps

Turning now to active transport; cells maintain concentration differences of ions across membranes by promoting their movement against concentration gradients. The hydrolysis of ATP provides the energy to drive this process, and proteins, which act as conduits, are known as pumps. Pumps that use ATP directly are referred to as primary active transport systems. As these proteins catalytically hydrolyse ATP, they are also referred to as **ATPases**.

Box 4.5   ATP – the energy currency of life

Adenosine triphosphate, ATP (**4.1**) is used to drive many biochemical reactions and cellular processes that require the input of energy.

adenine ring

triphosphate group

ribose ring

4.1

Energy is obtained from ATP when it is broken down to adenosine diphosphate (ADP) and free phosphate ions. This reaction is thermodynamically favourable, having a free energy change, $\Delta G_m^{\ominus}$, of about -40 kJ mol^{-1}.

Note that adenine bound to a ribose ring is often referred to as adenosine.

$$\text{adenosine} - O - \overset{\overset{\displaystyle O}{\|}}{\underset{\underset{\displaystyle O^-}{|}}{P}} - O - \overset{\overset{\displaystyle O}{\|}}{\underset{\underset{\displaystyle O^-}{|}}{P}} - O - \overset{\overset{\displaystyle O}{\|}}{\underset{\underset{\displaystyle O^-}{|}}{P}} - O^- + H_2O$$

ATP

$$\text{adenosine} - O - \overset{\overset{\displaystyle O}{\|}}{\underset{\underset{\displaystyle O^-}{|}}{P}} - O - \overset{\overset{\displaystyle O}{\|}}{\underset{\underset{\displaystyle O^-}{|}}{P}} - O^- + {}^- O - \overset{\overset{\displaystyle O}{\|}}{\underset{\underset{\displaystyle O^-}{|}}{P}} - OH + H^+$$

ADP

ATP is synthesised by the reverse process, i.e. the addition of phosphate to ADP. The energy input for this reaction comes from the breakdown of organic fuel molecules, such as glucose.

You have probably noticed the negative charge on both ATP and ADP. This is balanced in these, and other phosphate-containing biological molecules (e.g. DNA), by cations, usually Mg^{2+}. In fact, both ATP and DNA may be regarded as magnesium complexes.

The ion and voltage gradients created and maintained by the primary pumps are essential to the viability of a cell. They store energy which is used for many purposes, perhaps the most important of which is to create secondary ionic gradients. To function effectively, a cell needs to establish concentration differences across the membrane, for many different ions. Rather than have a primary pump to create each difference, the cell can exchange one concentration or voltage difference for another by means of secondary pumps or ion exchangers. These are proteins that span the membrane and use the work done by one type of ion when it crosses the membrane to power the transfer of another type of ion across the membrane. A protein that transports both the driving and the driven ion in the same direction across the membrane is known as a **symport** (or coport) whereas one that transports the driving and the driven ions in opposite directions is known as an **antiport** (or counterport) as illustrated in Figure 4.13. We will meet an example of this type of secondary active transport later in Section 4.6.2 in the transport of iron.

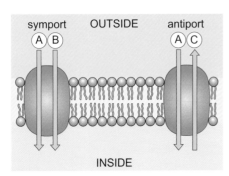

Figure 4.13 Schematic showing a symport and an antiport where A, B and C are different ions.

4.5.1 Sodium–potassium pump

The most studied primary pump is the sodium–potassium pump (Na^+/K^+-ATPase), which is present in the cell membrane of virtually all animal cells. From Table 4.1, you saw that the concentration of K^+ inside the cell is higher than outside whereas that of Na^+ is the opposite. These ionic gradients are maintained by the activity of this pump. During the hydrolysis of ATP, Na^+/K^+-ATPase transfers three Na^+ ions outwards across the membrane, and simultaneously transfers two K^+ ions inwards, the energy liberated when ATP is hydrolysed being sufficiently large to provide the energy required to move the cations against their respective electrochemical gradients. Thus the pump is creating an uneven charge distribution across the membrane; such a transporter is referred to as being **electrogenic**.

The protein responsible is a tetramer comprising of two membrane-spanning α-chains (which loop backwards and forwards through the membrane) and two smaller β-subunits, which are largely exposed to the outer surface and carry **oligosaccharide** chains. Chemical cross-links can be readily formed between the two α-subunits or between α and β, but not between β and β, suggesting that the α-chains are in contact but the βs are not (Figure 4.14).

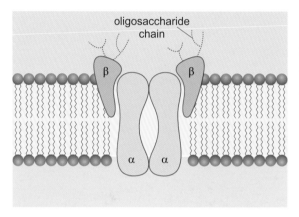

Figure 4.14   The structural organisation of Na^+/K^+-ATPase.

The spatial organisation of the ATPase in the membrane has been investigated using so-called erythrocyte 'ghosts'. If red blood cells (**erythrocytes**) are added to water they immediately burst; this is a consequence of **osmosis**, the movement of a solvent (usually water) from a region of high water concentration to one of low water concentration across a partially permeable membrane. The force that drives the solvent molecules through the membrane is called the **osmotic pressure**.

If erythrocytes are immersed in a **hypotonic** salt solution, that is one whose osmotic pressure is lower than that of the inside of the cell, they swell and become leaky. The haemoglobin is lost and a pale ghost cell is left with an internal composition in equilibrium with the surrounding medium. If the cells are then resuspended in an **isotonic** solution, that is one with the same osmotic pressure as the contents of the cell, the membrane reseals and again functions as a permeability barrier. In this way, the internal composition of the ghost can be varied. Incidentally, you may have heard the terms hypotonic and isotonic used in the context of sports drinks (Box 4.6).

Box 4.6 An aside: recharging your batteries after exercise – do sports drinks provide the solution?

To anyone who exercises regularly, rapid rehydration and replenishment of energy supplies following a workout are essential. During exercise, the body maintains a steady temperature of 37 °C by the production of sweat and subsequent cooling by evaporation; however, there is a drawback – in addition to loss of water, essential ions such as Na, K^+, Ca^{2+}, PO_4^{2-} and Cl^- are also lost. And that's not all, the muscles increase the uptake of glucose, and the body's carbohydrate stores are depleted. In this regard, it has long been known that there is a close correlation between blood sugar levels and the physical condition of athletes following intense exercise. In 1924, Samuel A. Levine, a medical practitioner, together with some colleagues, published a study (Levine et al., 1924) in the *Journal of the American Medical Association* on the composition of the blood of several runners who had completed the Boston marathon. Those with normal blood sugar levels were in 'excellent condition after the race' whereas those with highly reduced blood sugar levels were 'markedly prostrated at the finish'.

Simply drinking water after exercise is not sufficient to re-establish fluid balance; indeed ingesting large volumes of water only serves to increase urine production which is rather counterproductive, or may even kill you – there is recent evidence suggesting that 'water intoxication' may be responsible for the deaths of runners from too rapid ingestion of large volumes of water. A cocktail of carbohydrate and essential ions in aqueous solution would seem preferable. It is here that specially formulated sports drinks play a part. There are three types. Isotonic drinks have similar carbohydrate and sodium ion contents to the body's own fluids and are the most common choice for most athletes. Hypertonic drinks have higher carbohydrate levels, and are used to supplement daily carbohydrate intake for very long distance events such as marathons (although during exercise, athletes are advised to take them in conjunction with isotonic drinks to restore ideal fluid levels). In contrast hypotonic drinks have a low level of carbohydrate and are used to replace quickly fluids lost through sweating where a boost of carbohydrate is not required; these tend to be favoured by jockeys and gymnasts.

However, these drinks are not without their detractors, not least those who suggest there is no need to purchase expensive brands, and instead you can make your own from orange squash and a pinch of salt (or indeed a banana and a glass of water will work just as well), but whatever, it does appear that this is an area where chemistry can help sporting performance – and of course it's entirely legal.

4.5.2 Structure and function of the sodium–potassium pump

Experimental investigations have established that Na^+ and ATP must be present inside the cell, whereas K^+ must be outside to enable transport to occur.

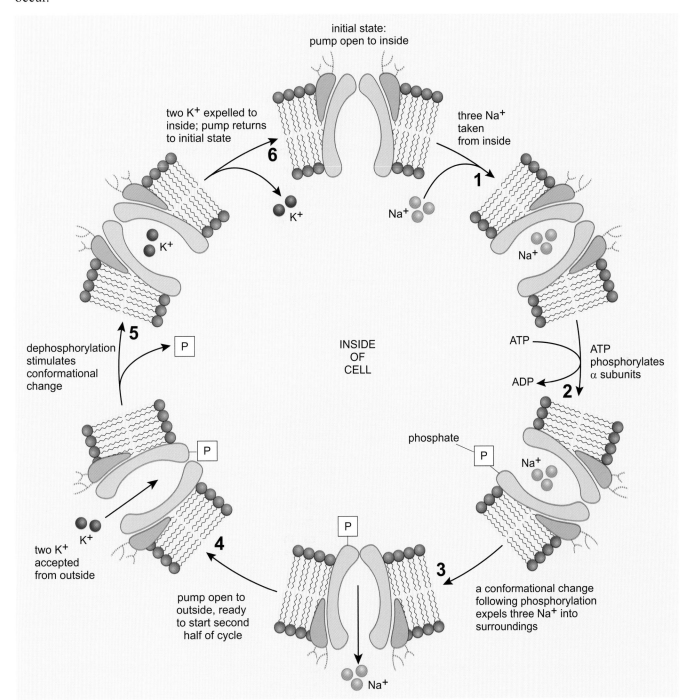

Figure 4.15 Schematic model of the sodium–potassium pump in operation. Steps of the pump's operation are shown as a cycle. (Note that Mg^{2+} – not shown here – is an additional cofactor required by all ATPases.)

ATP is not hydrolysed unless Na^+ and K^+ are transported; tight coupling with transport is achieved by the transient phosphorylation (by ATP) of an aspartate residue on the α-subunit, which occurs when Na^+ is present.

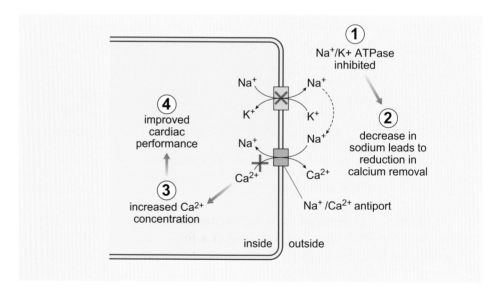

aspartate residue   phosphorylated
                                      aspartate residue

Pumps comprising the α- and β-subunits, where a polypeptide is phosphorylated during transport, are often termed P-type.

This promotes a conformational change of the protein between two states, causing the movement of Na^+ ions out of the cell. The phosphorylated aspartate residue is subsequently hydrolysed in the presence of extracellular K^+ ions that can now bind to a site exposed by the movement of Na^+ ions. K^+ ions are then transported to the interior of the cell as the ATPase returns to its original conformation.

A schematic diagram of the sodium/potassium pump in operation is shown in Figure 4.15. The pump is used to regulate the intracellular levels of these ions.

Figure 4.16 The effect of digitalis on the heart. (1) The Na^+/K^+-ATPase is inhibited, diminishing the Na^+ gradient across the cell membrane. (2) Ca^{2+} extrusion is coupled to Na^+ entry by a specific antiport. (3) The decreased Na^+ gradient leads to a decrease in the extrusion of Ca^{2+} from the cell. (4) The increased intracellular Ca^{2+} concentration improves the performance of heart muscle cells.

There are drugs that inhibit the dephosphorylation of ATP by interacting with a specific binding site on the protein pump, which is located on the extracellular face of the membrane. For example, the clinical benefits of digitalis, which is derived from the foxglove plant, have long been recognised in the treatment of heart failure. The drug inhibits the Na^+/K^+ pump, thus diminishing the Na^+ gradient which, in turn, leads to an increase in intracellular Ca^{2+} concentration, as Ca^{2+} is usually extruded by a Na^+/Ca^{2+} antiport. The elevated cytosolic Ca^{2+} concentration within the cell enhances cardiac performance (Figure 4.16). However, cases of poisoning by digitalis, both accidental and deliberate, are numerous. A purely fictional example of the latter may be found in Agatha Christie's novel *Appointment with Death*.

As for ion channels, X-ray crystallography has revealed much about the structure of ion pumps. Figure 4.17 shows the structure of one of the conformations of the sodium/potassium pump determined by Bente Vilsen and Poul Nissen from the University of Århus, Denmark, and published in the prestigious science journal *Nature* (Morth et al., 2007). As a consequence of the experimental approach used, the structure shown contains the MgF_4^{2-} ion, which mimics the phosphate group, and Rb^+ ions are incorporated in place of K^+. An additional protein subunit designated γ, often associated with α and β and which is thought to regulate the activity of the pump, is shown here. Structurally a close relationship between the metal binding sites in the sodium/potassium pump and those in Ca^{2+}-ATPase was found.

In contrast to animals, plants (and bacteria) do not possess a sodium–potassium pump, but instead use a proton pump to power the movement of ions across membranes, as we shall see in the next section.

4.6 Metal uptake – iron transport across a cell membrane

Before we move onto transport of metals ions around the body, we will return to consider the example of iron and, in particular, how it is taken up by an organism.

■ In Chapter 3 we discussed the different mechanisms that organisms use to aid the solubility of the iron, resulting in a form that can be absorbed or assimilated by the organism – what are the two main forms of soluble iron that we have considered?

☐ The iron is either complexed as Fe(III), for example in the Fe(III)–siderophore or Fe(III)–phytosiderophore complex, or is reduced to Fe^{2+}(aq), for example at the root surface in plants.

So how does an organism actually take up the iron? We will consider the different mechanisms in turn.

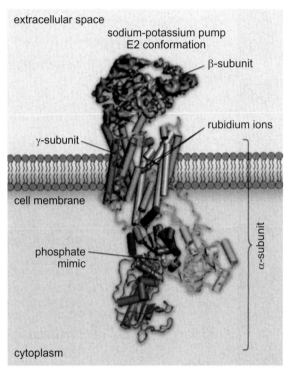

Figure 4.17 Structure of the sodium–potassium pump.

4.6.1   Iron(III)–chelate transporters

In Section 3.3.3 you met the idea that iron can be taken up as an Fe(III) complex, such as Fe(III)–enterobactin in *E. coli* – but how is this complex transported across the outer membrane into cells? First let us consider *E. coli* in more detail (Figure 4.18). In a schematic close-up of the *E. coli* membrane shown in Figure 4.18b, you can see that there are in fact two membranes separating the exterior of the cell and the cytosol: an outer membrane and a cytoplasmic or inner membrane. These are separated by the **periplasmic space**, an aqueous environment that resembles the environment outside the cell.

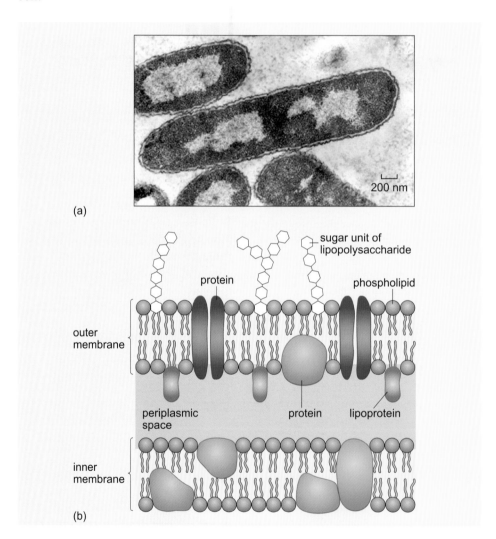

Figure 4.18   (a) *E. coli.* (b) Schematic showing a close-up of the structure of the cell membranes in *E. coli.*

The Fe(III)–siderophore complex is transported into the cell via several steps. *E. coli* possesses specific membrane receptor proteins for Fe(III)–siderophore complexes. An example of one of three outer membrane receptors whose crystal structures have recently been determined is shown in Figure 4.19a and b.

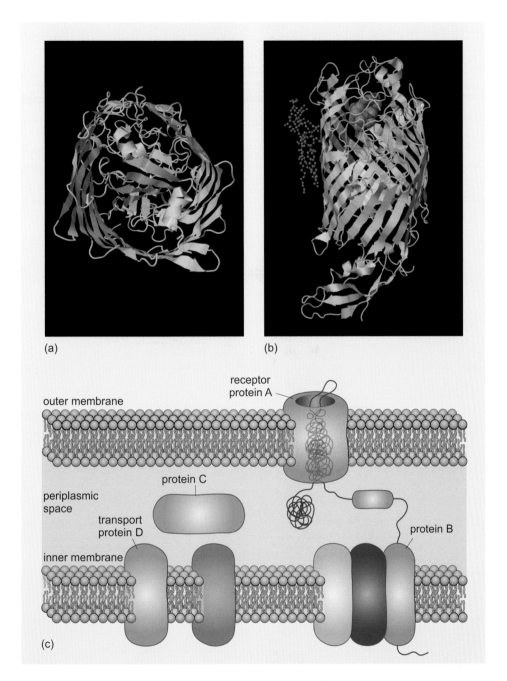

Figure 4.19 (a) Top view of the crystal structure of an outer membrane protein (protein A). (b) Side view of a Fe(III)–siderophore complex (shown as a space-filled representation) bound to an outer membrane receptor (protein A). A molecule of a lipopolysaccharide is shown associated with the outer membrane protein. The chain in pink represents protein B. (c) Schematic model of the transport of Fe(III)–siderophore across the outer and cytoplasmic membranes of *E. coli*. (Based on the following pdb files: (a) 1fep (Buchanan et al., 1999), (b) 2grx (Pawelek et al., 2006).)

These receptors contain a β-barrel (where the β-sheet has rolled into a barrel shape) that spans the outer membrane. There are also segments inside the barrel which effectively cork the barrel from the periplasmic side of the outer membrane. A schematic representation of the transport mechanism is shown in Figure 4.19c although the exact mechanism is not fully understood. The Fe(III)–siderophore complex is believed to bind to the outer membrane receptor protein (A) and is transported into the cell through interaction with a second protein (B) spanning the periplasmic space which, through interaction with its adjacent proteins, triggers a conformational change in the receptor protein (A).

Once released into the periplasmic space, the Fe(III)–siderophore complex is bound by another protein (C), preventing reverse transport back across the outer membrane. A variety of transport proteins (D) mediate diffusion across the cytoplasmic membrane. Once in the cytosol, the iron is then released for use by other proteins as discussed in Section 3.3.4.

A similar mechanism is likely to occur in plants, where the Fe(III)–phytosiderophore complexes are transported into the cytoplasm via a highly specific transporter. The chelating ligand is separated on reduction of the metal.

4.6.2 Divalent ion transporters

We will now consider the uptake of Fe^{2+}(aq) and other divalent metal ions looking initially at plants. Plants use a proton pump (H^+-ATPase) to power the movement of ions across membranes. They do this by generating a proton gradient across the membrane. In root cells, dissipation of this proton gradient is coupled to the uptake of metal ions from the soil. The M^{2+} ions are transported into the plant by a divalent ion transporter, IRT1, as shown in Figure 4.20a.

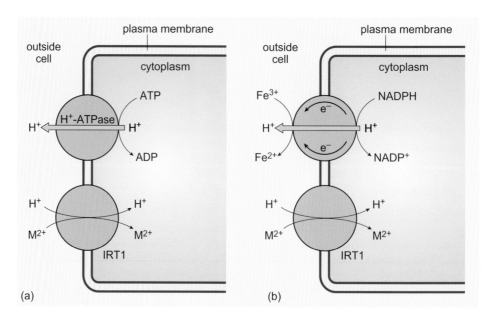

Figure 4.20 Transport of divalent metal ions across the cell membrane in plants. NADPH and NADP$^+$ are the reduced and non-reduced forms of nicotinamide adenosine dinucleotide phosphate, a coenzyme that can provide H$^+$ and electrons.

You saw in Section 3.4, in the constitutive mechanism in plants, that the uptake of iron from soil is aided by secretion of both H^+ and a reducing agent from the roots. In this case there is evidence that the proton pump may be linked to an electron transport chain in the plasma membrane, together referred to as a transmembrane redox pump, shown in Figure 4.20b. $Fe^{3+}(aq)$ is reduced to $Fe^{2+}(aq)$ which can be taken up by the roots by the divalent transporter.

■ What type of transport is illustrated in Figure 4.20?

☐ This is an example of secondary active transport and IRTl is an example of a $M^{2+}-H^+$ symporter.

A divalent ion transporter is also believed to be important in the uptake of iron in mammals. Iron from digested food is absorbed in the duodenum by the **enterocyte** cells that line the small intestine. Iron is transported into these cells as the $Fe^{2+}(aq)$ ion, a ferric reductase enzyme present in intestines reducing any $Fe^{3+}(aq)$ to $Fe^{2+}(aq)$. The $Fe^{2+}(aq)$ is transported across the cell membrane of the enterocyte by a protein called **divalent metal transporter DMT1**. This transporter is another example of a $M^{2+}-H^+$ symporter using the concentration gradient of protons (from gastric acid) and the electrical potential of the membrane to drive the uptake of $Fe^{2+}(aq)$.

> It is the requirement for protons for cotransport of the iron that explains why antacids can interfere with iron absorption and why iron absorption is increased if you take iron supplements, for example, with orange juice.

Both the divalent ion transporters DMT1 and IRT1 are integral membrane proteins, with several transmembrane domains that transport Fe^{2+} directly across the membrane. The iron is either stored in the enterocytes (as ferritin) as we shall see in Chapter 5 or is transported out of these cells into the blood stream by another protein, the iron exporter **ferroportin**. Once in the bloodstream the iron is oxidised to Fe(III) by another protein called **hephaestin**, a copper-dependent ferroxidase, ready for take up by another transport protein as we will see in Section 4.7.1. Figure 4.21 summarises these important pathways for iron uptake in mammals.

■ The intestines of animals and the rhizosphere of plants (the region of soil directly surrounding the roots) are anaerobic environments with limited oxygen present. Why are these divalent metal transporters unlikely to be important in oxygen-rich environments?

☐ In oxygen-rich conditions, Fe(III) will predominate. In anaerobic environments such as the intestines, it is energetically more favourable to transport the divalent ion.

DMT1 is a non-specific divalent ion transporter. Other transporters exist, some specific to particular metal ions. For example, copper has two specific transporters, known as CRT1 and CRT3. Like iron, copper chemistry poses difficulties for its uptake, its solubility depending on oxidation state and chelators present. In copper, it is the Cu(I) species that appears to be the preferred species for uptake. $Cu^{2+}(aq)$, which is the stable species in aerobic conditions, is reduced to $Cu^+(aq)$ by the same reductases responsible for the reduction of $Fe^{3+}(aq)$. $Cu^+(aq)$ is then transported across cell membranes by the transporters CRT1 and CRT3.

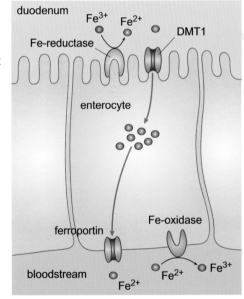

Figure 4.21 Iron transport across an intestinal enterocyte.

■ Considering the potential diagram for copper, shown below, what possible difficulty can you see that arises for the transport of Cu^+?

$$Cu^{2+} \xrightarrow{\text{0.16 V}} Cu^+ \xrightarrow{\text{0.52 V}} Cu$$
$$\xrightarrow{\text{0.34 V}}$$

☐ Free Cu^+ will undergo disproportionation to Cu^{2+} and Cu^0 in aqueous solution.

Cu^+ however is bound by many chelators, has a faster ligand exchange rate and is believed to provide a more specific transport system than Cu^{2+}.

Box 4.7 Vanadium enrichment in sea squirts

You saw in Table 3.1 that certain low abundance metals have higher concentrations within a cell than are found in seawater – so how does this enrichment occur?

Figure 4.22 Sea squirts.

Let's consider the example of vanadium, which is found to be present at high levels (at least four orders of magnitude compared to seawater) in the blood cells of certain types of sea squirts (vanadytes). The exact function of the vanadium in these sea squirts (shown in Figure 4.22) is not known, although there is some debate that the vanadium might play a role as an oxygen carrier in the blood. Clearly sea squirts have a mechanism for accumulation of vanadium from seawater. Vanadium exists in seawater as the vanadate ion, VO_4^{3-}. This anion enters the blood cell by means of a phosphate channel, known to transport anions through the outer membrane of a cell. Once inside, the vanadium is reduced and accumulates within the cell, either in the cytoplasm or in vacuoles. We will look at the storage of metal ions in cells in Chapter 5.

4.7   Metal transport around the body

So far, we have only discussed how metals are transported into a cell; however considering the example of iron in humans, it is obvious that metals must be transported around the body. (Plants also have a requirement for metal transport – see Box 4.8.) Metals, from our diet, are absorbed into the bloodstream and transported from food in the gut to the parts of the body where they are required. The alkali metal ions (such as Na^+ and K^+) and, to a lesser extent, the alkaline-earth metal ions (Ca^{2+} and Mg^{2+}) are soluble and can be transported as free ions in the bloodstream. At relatively high concentrations compared to the trace elements, they do not require specific mechanisms for transport, as they interact only weakly with ligands in the bloodstream. The transition metal ions, however, require specific carrier proteins in the bloodstream to transport the ion, as shown for a number of ions in Table 4.3. We will start once again by looking at the example of iron.

Table 4.3   Transport proteins present in blood plasma.

Ion	Transport protein
Fe^{2+}	transferrin
Cu^{2+}	ceruloplasmin, albumin
Zn^{2+}	albumin
Ca^{2+}	phosphoproteins
Na^+, K^+, Mg^{2+}	none required

As we will see the transport of iron presents its own problems. As bacteria secrete powerful chelators such as siderophores, iron must be kept under very close control. Any free iron within an organism is available for chelation by a siderophore, which may lead to bacterial infection within the organism. We shall examine the biochemical systems that handle iron within the human body in the next section.

Box 4.8   Metal transport in plants

Metals must also be transported in plants, from the roots to the shoots. The transport of micronutrients in plants involves the **phloem**, a circulatory system in plants, which uses the movement of sap from cell to cell. To be transported, the metal cations clearly need to be soluble in the sap, which has a pH of 7.5–8.5. As in mammals, the transition metal cations are transported as complexes. A number of naturally occurring ligands have been proposed, including amino acids, di- and tri-carboxylic acids, amides and amines, in particular nicotianamine, **4.2**.

4.2

4.7.1 Iron transport in the blood – transferrin

Iron, for the most part, is required in the bone marrow, where the red blood cells are formed. The iron in red blood cells is carried around the body in haemoglobin, which we will consider in detail in Chapter 8. Red blood cells have a finite lifetime of about only four months, and old cells are destroyed, usually in the spleen. Iron from the destruction of these cells is then transported from the spleen back to the bone marrow to be recycled. (Most of the iron in the body is stored and recycled; however, humans lose an estimated amount of 1 mg per day on average from sweat and shedding of skin cells, as well as loss of blood, hence the requirement for a daily intake of iron. This is particularly important in women of child-bearing age because of menstrual loss.)

Iron cannot be transported around the body's circulation system as free iron as discussed in Section 3.2.1, because it would be susceptible to chelation by siderophores, may precipitate as iron(III) oxide, or may form iron(II), promoting the formation of free radicals. Therefore, a specific transport protein is required, called **transferrin**. (In fact, a whole class of transferrin-like proteins is involved in iron transport.) Transferrin is a medium-sized protein with a relative molecular mass of about 80 000. The crystal structures of transferrin with and without iron have been obtained, and the overall structure is shown in Figure 4.23. The structure of apo-transferrin without any iron (Figure 4.23a) shows that transferrin can be considered as two very similar polypeptides back-to-back, with each of the polypeptides having a large cleft.

The apex of each cleft coordinates one iron atom; each transferrin molecule is therefore capable of transporting two iron atoms. On the binding of iron, there is a significant change in the higher-order structure of the protein, such that the two sides of the cleft come together and incarcerate the iron atoms (Figure 4.23b). Both iron atoms are now buried deep within the protein structure. (It is not fully clear why the iron atoms are buried in this way, but it may help in protecting the iron atom from microbial siderophores.) Release of iron from transferrin is accompanied by a subsequent rotation of the domains.

The iron binding site in transferrin is rich in hard ligands, which are suitable for binding iron(III) in a stable complex. When the iron atom enters the active site (Figure 4.24), it is coordinated by one aspartyl, one histidyl and two tyrosinate side chains; a non-protein ligand also coordinates to it. This external ligand is a carbonate, CO_3^{2-}, which is held in place within the protein via hydrogen bonds to the protein backbone.

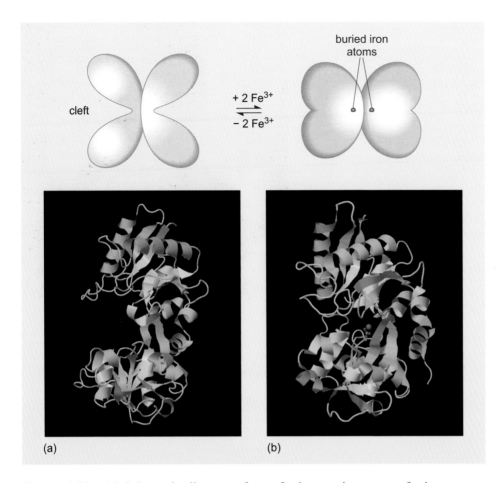

Figure 4.23 (a) Schematic diagram of transferrin protein apotransferrin; (b) proposed higher order structure change on complexation of iron(III) showing the hinge motion associated with the iron carbonate binding. The images at the bottom show just one of the polypeptide chains. (Based on the following pdb files: (a) 1tfa (Mizutani et al., 1999), (b) 1d3k (Yang et al., 2000).)

■ What is the mode of bonding of the CO_3^{2-} group?

☐ The carbonate coordinates to the iron in a bidentate fashion; in other words, it is a chelating ligand.

Once the carbonate is held in place, we can see another example of ligand preorganisation, where an octahedral environment of hard/borderline ligands is ready to receive the iron(III) ion. The carbonate binding appears to facilitate the iron binding by the protein and vice versa, and so the system is said to be **synergistic**. It is not clear why this unusual synergistic binding of iron and carbonate occurs in transferrin, but it may have something to do with the way iron is released from transferrin.

As all the coordinating ligands in transferrin can be considered as hard or 'borderline' ligands, it is no surprise that transferrin forms a very stable complex with iron(III). The stability constant of the Fe(III)–transferrin complex is about 10^{20} mol^{-1} dm^3. This is high enough to protect the iron(III) against the low concentration of any siderophores present. Indeed, the

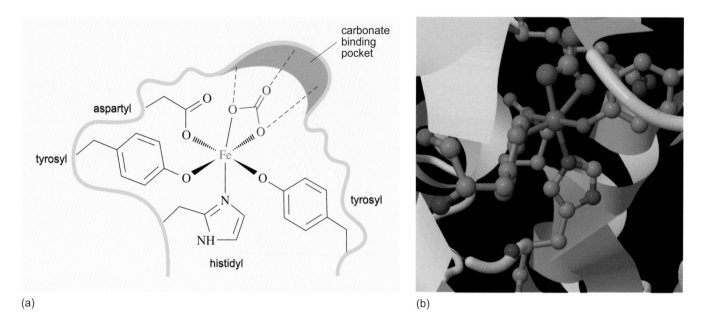

(a)   (b)

Figure 4.24 (a) Schematic and (b) crystal structure of the six-coordinate iron binding site in transferrin; the coordination geometry is distorted octahedral. The carbonate is held in place by hydrogen bonds (blue lines) to amino acid residues in the protein backbone. ((b) Based on pdb file 1d3k (Yang et al., 2000).)

transferrins show mild antibacterial properties, in which their method of operation is to prevent extensive iron chelation by siderophores (see Box 4.9).

Box 4.9 Iron in human milk

It has been known for some time that bottle-fed babies are more likely to suffer from gastric infections than breast-fed babies; this may be despite strict hygiene standards. The reason for this is thought to lie in the availability of iron within the baby's feed. Breast milk is known to contain a transferrin-like protein called lactoferrin. The lactoferrin chelates all the iron in the mother's milk, and prevents iron chelation by microbial siderophores. Formula milk, on the other hand, does not contain human lactoferrin, so the iron in the feed is more available for chelation by siderophores secreted by bacteria.

Transferrin also forms relatively stable complexes with other hard metals (Table 4.4). Current thinking suggests that these other metals are transported by transferrin into cells, where they are potentially detrimental.

Table 4.4 Stability constants of metal–transferrin complexes.

Metal	Log (stability constant)
cadmium(II)	5.95
zinc(II)	7.8
aluminium(III)	13.5
iron(III)	22.8

Therefore, we can see from the structure and function of transferrin that the transport of iron within the body is very carefully managed, so as not to allow any free soluble iron to form.

The transferrin transport protein is present in the bloodstream; however, iron is obviously required at a cellular level. So, how is the iron transferred from transferrin into a cell and how, then, is iron stored? After all, we must store iron, as we need a reservoir of it for the synthesis of iron-containing proteins, most notably haemoglobin and myoglobin. As with iron transport, the iron storage systems need to ensure that free, soluble iron is not formed. We will consider the question of storage in Chapter 5, but first will consider how the iron is removed from transferrin.

4.7.2 Removal of iron from transferrin

Like enterobactin, uptake of the iron-containing transferrin complex into a cell occurs via binding to a receptor on the surface of the cell membrane. However, in this case, as shown schematically in Figure 4.25, the part of the membrane containing the receptor and transferrin breaks off inside the cell to form a vesicle in a process called **endocytosis**.

The vesicle is coated with a protein called clathrin, which comes off once inside the cell. The pH in the vesicle then decreases to about 5.5 as a membrane-bound enzyme (H^+-ATPase) pumps protons into the vesicle.

- ■ What will be the effect of a decrease in pH on the stability of the transferrin complex?

- □ A decrease in pH is likely to decrease the stability of the complex. Some of the ligands binding the iron, including possibly the carbonate, will become protonated, facilitating release of iron from the transferrin.

Research suggests that, in particular, it is protonation of two lysine residues in one of the lobes of transferrin that aids release, as strong electrostatic repulsion of the positively charged lysines when protonated drives the domains apart.

Once inside the endocytic compartment (the endosome), the iron is reduced to Fe^{2+} making it available to DMT1, the divalent metal transporter, and is transported into the cytoplasm within the cell. It is not clear whether the

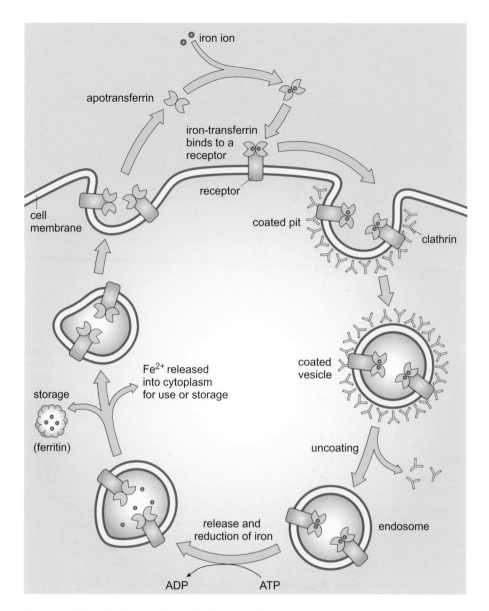

Figure 4.25 Endocytosis cycle for transferrin.

reduction is part of the release mechanism or occurs subsequently. The reduction potential of Fe^{3+} bound to free transferrin is too low to be reduced physiologically, even at this lower pH; however, binding to the receptor raises the potential so that reduction is now possible. As with many other complexes, Fe^{2+} is bound to transferrin much more weakly than Fe^{3+}.

The vesicle containing the iron-free apotransferrin then fuses with the outer membrane, re-exposing the apotransferrin to the outside of the cell where it is released and is free to bind more iron. Cells throughout the body have such receptors for iron-containing transferrin complexes and so obtain iron from the circulating blood.

■ Which of the three main strategies for iron mobilisation and acquisition does this process involve?

□ The uptake of iron into cells from the bloodstream combines all three: chelation, acidification and reduction.

Interestingly a genetic abnormality has been identified in patients suffering from haemochromatosis (or iron overload), in a protein designated HFE. In its normal state, HFE associates with the transferrin receptor and suppresses iron uptake. However, in a mutated form associated with the abnormality, there is a loss of association and so greater iron absorption occurs.

4.8 Intracellular transport

We will now turn our attention to the transport of metals within a cell itself. Once a metal has been transported into a cell, it can be used, for example for the production of enzymes, but how does it 'find' the correct protein or enzyme? Traditionally it has been thought that this is controlled by the specific chelating properties of the individual apoproteins, in particular the side chains providing a particular coordination or geometry suited to metal ions of a particular ionic radius or electronic preference.

■ What difficulties can you see with this particular scenario?

□ The ionic radii and coordination preferences of many of the metal ions are similar. In addition, many of these enzymes are large and the active sites may be buried deep within the protein.

Another possibility is the use of **metallochaperones**. We will consider now the example of copper, which is required in several different enzymes. Given the low levels of Cu found in the cytoplasm, there is evidence that a cell uses specific proteins called metallochaperones to acquire, transport and deliver Cu to where it is required. A metallochaperone binds the Cu (as Cu(I)) after it enters the cell and transports it, protecting it from intracellular copper chelators, before donating it to the enzyme that needs it. The metallochaperone is believed to facilitate insertion of the copper into the enzyme. Two main classes of copper metallochaperones have been identified in eukaryotes, each transferring Cu to a specific target enzyme. The small soluble metallochaperone molecules bear a close resemblance to the amino acid sequence and structure of the binding site of the target enzyme.

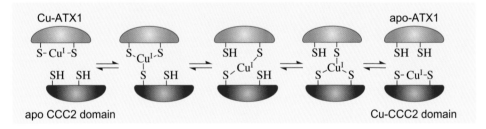

Figure 4.26 A copper transfer mechanism for the ATX1 copper chaperone and a copper-transporting ATPase (CCC2).

Figure 4.26 shows the proposed 'docking' mechanism for the ATX1 chaperone (Cu-ATX1) and its target, a copper-transporting ATPase (CCC2).

This metallochaperone was initially identified in yeast (*S. cerevisiae*) but has since been found in plants, insects and humans. So far, most research on metallochaperones has been restricted to copper; however, given the infancy of this field of research (metallochaperones were unknown until 1997), it is possible that other chaperone molecules may exist for other metals within the cell and have not yet been identified.

Don't forget that there are questions on the companion website which you can use to test your understanding of the material covered in this chapter.

5 Metal storage

In the previous two chapters, we considered how metals are acquired from the surrounding environment (or, for humans, from our diet; Chapter 3) and how they are transported into cells and around the body (Chapter 4). Now, a high proportion of the metal ions in the body will be 'tied up' in metalloproteins, the blood and of course bone (Box 5.1), which can act as a store of calcium. Obviously, the levels of metal ion nutrients can vary, as can the demand for a particular metal. How, then, are these metal ions stored, for example in times when they are abundant, until they are required? In addition, when high concentrations of metal ions are accumulated within a cell these ions can become toxic, so how can they be prevented from attaining toxic levels due to metal overload? In this chapter we will look at some of the mechanisms living systems use to store metal ions. In particular we will concentrate on storage within cells and the ways that cells can combat high metal ion concentrations.

Cells use two main strategies for storing metal ions. In the first, the metal is bound by proteins or macromolecules present within the cytoplasm, thus decreasing the concentration of free metal ions available. In the second, the metal is transported into membrane-bound compartments within the cytoplasm, such as vacuoles in plants, where separated by a membrane, the metal is less damaging. We will meet both these strategies in the following sections.

Box 5.1 Bones as a calcium store

Bone, as we will see in Chapter 6, is a biomineral consisting of hydroxyapatite, $Ca_{10}(PO_4)_6(OH)_2$ and is a depository for calcium, accounting for approximately 99% of the Ca in an average person. Bone is continually being deposited and reabsorbed into the bloodstream by cells known as osteoclasts and osteoplasts. When our diet lacks sufficient calcium for our needs, common for example in pregnancy, the deficiency can be met from our bone. This loss of calcium is harder to replace as we get older, particularly for women after the menopause and may lead to osteoporosis.

5.1 Iron storage in the human body: ferritin

Once again, we will initially focus our discussion on iron. Humans clearly need a reservoir of iron in the body for the synthesis of iron-containing proteins, most notably haemoglobin and myoglobin. As with iron transport, the iron storage systems need to ensure that free, soluble iron is not formed.

In humans, iron is stored mainly in the bone marrow (where haem is synthesised), spleen and liver. About 10% of all the iron in the body is in storage. (The majority of the iron present in the body is of course circulating in the bloodstream.) Two proteins are involved in iron storage; these are called **ferritin** and **haemosiderin** (they also occur in other organisms). Here, we shall only study the better characterised (and simpler) ferritin which is found in the liver.

Each ferritin molecule can store iron up to about 20% of its total mass. This is a very high percentage, considering that less than 0.2% of the total mass of proteins such as transferrin and myoglobin is iron. Ferritin is a large protein with a relative molecular mass of 440 000. The crystal structure of ferritin with no iron (apoferritin) is shown in Figure 5.1.

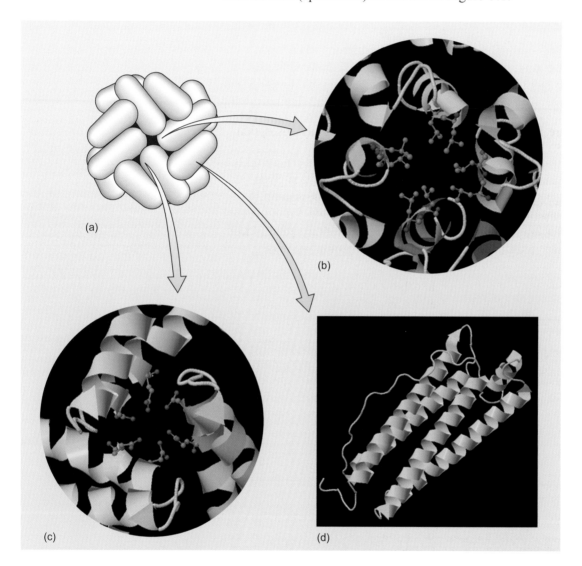

Figure 5.1   (a) Subunit assembly of apoferritin. Each subunit (shaped like a sausage) is made up of four parallel, α-helical polypeptide chains as shown in (d). (b) A fourfold channel and (c) a threefold channel leading to the centre of the structure are clearly visible. ((b)–(d) based on pdb file 1fha (Lawson et al., 1991).)

The overall structure shows that ferritin is a huge, hollow protein, with a wall mostly made up of α-helical peptide chains. These chains pack together to form a hollow sphere of about 8 nm in diameter. The structure is quite symmetrical, being roughly dodecahedral, and is one of the outstanding examples of symmetry in chemistry. The wall contains channels that lead from the inside to the outside of the hollow 'sphere'. There are two types of channel, a fourfold channel which forms at the intersection of four α-helical peptide chains and a threefold channel which forms at the intersection of three peptide chains.

■ The threefold channel is lined with aspartate and glutamate side chains and the fourfold with leucine – what does this tell you about the polarity (and hydrophilicity) of the different channels?

☐ Aspartate and glutamate are polar side chains and so the threefold channel will be hydrophilic, while leucine is non-polar and the fourfold channel will be hydrophobic.

5.1.1 Storage of iron in ferritin

The crystal structure of iron-containing ferritin is not known. However, some clues as to its structure have been obtained from extended X-ray absorption fine structure (EXAFS) studies. EXAFS gives information about the direct coordination environment of a particular atom in terms of the number and type of its coordinated atoms (although no angular information is usually available). EXAFS studies on iron-containing ferritin showed that each iron atom is surrounded by an inner shell of six or seven oxygen atoms at a distance of about 160 pm, and by a second shell of seven or eight iron atoms at a distance of about 290 pm (Figure 5.2). This was a very strange result. How could seven or eight iron atoms be packed around each iron atom? The problem was solved when it was noticed that the EXAFS data were very similar to that of a hydrated iron(III) oxide mineral called ferrihydrite, $5Fe_2O_3.9H_2O$. From this result, it was clear that ferritin stored iron partly as a crystalline, hydrated iron(III) oxide. Further studies showed that the inorganic crystalline part was within the hollow sphere of the protein.

Therefore, ferritin stores iron as crystalline, hydrated iron(III) oxide within its structure.

■ How will this affect the availability of iron?

☐ As iron(III) oxide is very insoluble, it is unavailable to microbial iron-chelating ligands.

The inorganic iron(III) oxide core is also protected from chelators by the outer protein coat. Moreover, this is a very space-efficient method of storing iron; each ferritin protein macromolecule can store a maximum of 4500 iron atoms. It is somewhat ironic that the formation of highly insoluble iron(III) oxides, which reduces the concentration of available iron in the environment, is the method by which iron is stored in ferritin.

Exactly how iron 'grows' within ferritin is not completely understood. Iron, as Fe^{2+}, is delivered to ferritin (after having been transported by transferrin), where it migrates through the carboxylate-rich channels in the surface of the protein to the interior. The inner side of the protein sphere is also rich in carboxylate residues. It is thought that these carboxylate residues coordinate iron atoms at a so-called ferroxidase centre, where Fe(II) is oxidised to Fe(III). An iron(III) oxide phase then grows within the ferritin core.

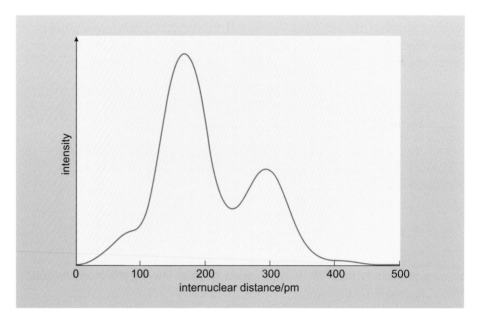

Figure 5.2 Fe–EXAFS radial distribution plot of iron-containing ferritin. Notice that there are two peaks, the first at 160 pm corresponding to a sphere of oxygen atoms, and the second at 290 pm corresponding to a sphere of iron atoms. (The peak due to the iron atoms is smaller than the peak due to the oxygen atoms; this is not in accord with the relative number of electrons in oxygen and iron atoms. The reasons for this are complex, but involve other factors beside the number of electrons in the intensity of backscattering. Also the actual structure of the hydrated iron(III) oxide in ferritin is not 'perfect', in that there are incomplete 'shells' of iron atoms, and poor long-range crystal order.)

■ What do we call this kind of process?

□ The growth of iron(III) oxide in ferritin is an example of biomineralisation – we shall study this process in more detail in Chapter 6.

5.1.2 Release of iron from ferritin

The exact mechanism by which iron is released from ferritin when required for use is again unclear. Overall the process must involve hydration of the iron(III) oxide and reduction of Fe^{3+} to Fe^{2+}, in a process that is essentially the reverse of mineralisation.

■ What possible difficulty can you see that this process entails?

□ This process involves dissolution of the solid mineralised iron core in ferritin and hence a change in phase from solid to aqueous ions. This is not a common reaction in biology.

Two main hypotheses currently exist for the release of iron from ferritin. These may not be mutually exclusive, occurring under different conditions or perhaps even simultaneously. One possibility is that release of iron may occur via gating or unfolding of the ferritin pores as shown schematically in Figure 5.3. Reductants such as NAD^+ (nicotinamide adenine dinucleotide) or

NADH (reduced NAD^+) present in the cell may enter via the open pores releasing the iron by a series of redox reactions with enzyme-catalysed reduction of Fe^{3+} to Fe^{2+}. After reduction, the iron is solvated and is believed to leave the ferritin core as the hydrated ion $(Fe(H_2O)_6{}^{2+})$.

■ Given that the iron is solvated after reduction, through which channels might it leave the ferritin core?

☐ The hydrated Fe^{2+} is likely to leave ferritin via the hydrophilic threefold channels. (The reductants are believed to enter via the hydrophobic fourfold channels.)

The other hypothesis is that ferritin is compartmentalised inside an acidic cytoplasmic vesicle, a lysosome, where the protein is digested, dissolving the mineral and releasing the iron.

Whatever process is used for solubilisation of iron, the organism must rapidly capture the free Fe^{2+} and transport it to the sites where it is to be utilised, therefore iron transport mechanisms must be activated at the same time as the signal for iron release is made.

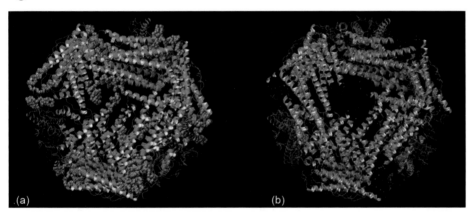

Figure 5.3 Gated ferritin pores. (a) Closed gates with pore helices. (b) Open gates, where pore helices are unfolded.

5.2 Control of iron uptake, transport and storage

We have seen that all organisms from humans, bacteria and plants need iron and so have the ability to regulate iron uptake and transport as required but what mechanisms exist to control this?

It is believed that iron may influence the mechanisms that genes use for synthesis of the proteins involved in iron uptake, transport and storage. The iron levels present in the cell may influence either gene transcription (using DNA as a target in the synthesis of messenger ribonucleic acid (mRNA)) and/ or translation (targetting the mRNA to regulate protein synthesis) (Box 5.2).

Box 5.2 The function of DNA and mRNA

The amino acids that form proteins are joined together with the help of ribosomes (sites of protein synthesis) in the cytosol of the cell. The mechanism by which the information encoded in the DNA (in the nucleus) is brought together with the ribosomes involves an intermediary class of molecules, ribonucleic acid (RNA). RNA carries the message from a template strand in the DNA to the ribosome, hence the name messenger RNA (mRNA).

Unlike DNA, mRNA molecules are not present in the nucleus all the time, but are made specifically to order using a short section of DNA as a template. Information within a cell can be seen as passing from DNA, via RNA, to the protein. There are two steps in this information flow – from DNA to RNA and from RNA to protein; these are called transcription and translation, respectively. Transcription is the process of synthesising RNA on a DNA template. The newly synthesised mRNA molecule is transported out of the nucleus, across the membrane with the aid of specialised and specific transporter proteins. Once in the cytosol, ribosomes become attached to the mRNA and the process of translation into a protein begins (Figure 5.4).

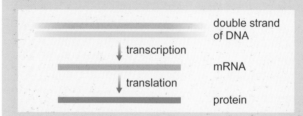

Figure 5.4 Information flow from DNA to mRNA to protein.

Considering first the synthesis of siderophores by bacteria, the genes involved in transcription are controlled by iron. A protein called FUR (ferric (iron) uptake regulator) appears to bind to specific segments of DNA in the presence of divalent ions such as Fe^{2+} but not in their absence. This binding represses transcription of the relevant genes involved in siderophore synthesis. In the absence of iron or at low metal ion concentration, the FUR protein does not bind to the DNA and gene transcription to produce siderophores can occur (Figure 5.5a).

In the synthesis of ferritin and the receptors that bind transferrin, control of the mRNA target that regulates translation or protein synthesis is believed to be important. Sequences called **iron-responsive elements (IREs)** exist within the mRNA sequences that are responsible for the post-transcriptional coding of transferrin receptors and ferritin. A binding protein, known as an **iron regulatory protein (IRP)**, binds to these IRE sequences in the absence of iron. In the synthesis of the transferrin receptors (Figure 5.5b), the IRP binds to the IREs on the transferrin receptor mRNA. This stabilises the mRNA and

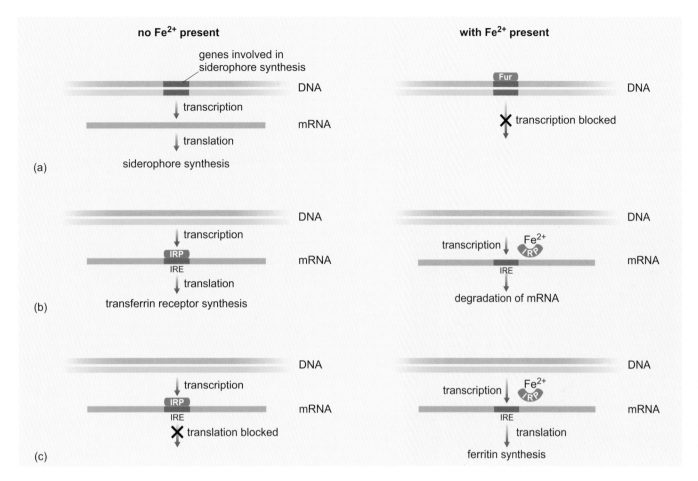

Figure 5.5 Regulation of transcription and translation for the synthesis of (a) siderophores, (b) transferrin receptors and (c) ferritin.

hence translation of the mRNA and the production of transferrin receptors can occur. In the presence of iron, however, the iron binds to the IRP causing it to change shape and unbind from the transferrin receptor mRNA. This transferrin receptor mRNA is rapidly degraded without the IRP bound to it.

■ How will this influence the production of transferrin receptors?

□ In the presence of iron, rapid degradation of the transferrin receptor mRNA will cause the cell to stop producing transferrin receptors and so the cell will not take up more iron.

In the synthesis of ferritin, an opposite effect with iron is observed (Figure 5.5c). When iron is present, the iron binds to IRP, again causing it to change shape so that it no longer binds to the ferritin mRNA. This now frees the mRNA to direct the cell to produce ferritin and hence increase storage of iron. One of the IRPs that has been identified incorporates an iron–sulfur cluster (4Fe–4S) when iron is abundant, but disassembles when iron is scarce. We shall meet other examples of Fe–S clusters in Chapter 7.

5.3 Metallothioneins

The **metallothioneins** are another class of storage protein present in the cytosol, which have been identified in many different organisms, including plants and mammals. In humans, they are found particularly within the kidney and liver. They are cysteine-rich proteins that are able to bind metal ions as metal–thiolate clusters.

■ Which metal ions would you expect to bind to cysteine ligands?

□ The cysteine side chain contains S, a soft ligand which would be expected to bind to soft metal ions such as Cu^+, Ag^+, Cd^{2+} and Hg^{2+}.

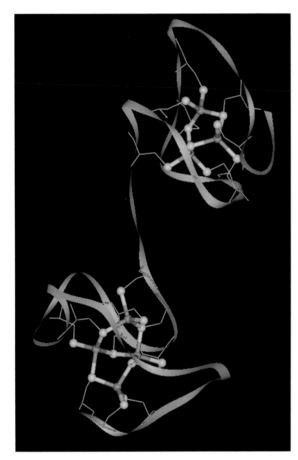

Metallothioneins bind these and other metal ions (including Zn^{2+}, Cu^{2+} and Pb^{2+}) with high affinities. However, some of these complexes are kinetically labile, for example that with Zn^{2+}, so that the metals are readily available for donation to other proteins. The metallothionein can thus be thought of as a 'store' for the metal ions.

The structure of the metallothionein proteins have been determined by X-ray crystallography and NMR, and an example is shown in Figure 5.6. They are low molecular weight proteins and typically bind between four and eight metal ions in small clusters. Although the actual amino acid sequences will differ for different metallothioneins, they have a similar structure, with two separate domains containing the metal–thiolate clusters built up of several metal–cysteine units.

■ What is the coordination of the metal in the metal–thiolate clusters shown in Figure 5.6?

□ The metal ions are in a tetrahedral coordination.

Metallothioneins may also serve a protective role, produced in large quantities when excess concentrations of metal ions, including toxic ones such as Cd^{2+} and Pb^{2+}, are present in the cell. Their synthesis, as in the examples for iron in Section 5.2 is controlled by the levels of metal ion present in the cell.

■ Which of the two strategies for metal storage within a cell are used by ferritin and metallothioneins?

□ Both ferritin and the metallothioneins are examples of metal-binding proteins present within the cytoplasm.

Figure 5.6 Structure of metallothionein determined by X-ray diffraction studies. The yellow atoms shown are sulfur atoms from cysteine residues and the blue atoms, zinc and cadmium ions. (Based on pdb file 4mt2 (Braun et al., 1992).)

5.4 Membrane-bound storage

We will now briefly consider the storage of metals (illustrated with the particular example of zinc) in membrane-bound compartments or organelles within the cytoplasm. Zinc is an important metal, found in many proteins (it has been estimated recently that as many as 10% of the approximately 30 000 different proteins in humans require zinc for their function). Zinc however can be toxic to cells if accumulated in excess and so there is a need for homeostatic control of the metal (as we saw for iron in Chapter 1). Much of the zinc in cells is bound by proteins in the cytoplasm as we saw in the previous section for the example of metallothionein. In addition, Zn^{2+} can be stored within membrane-bound compartments in the cytoplasm, thereby rendering the metal less damaging.

A reasonably well-understood example is the storage of zinc in the vacuoles in yeast. Uptake of zinc across the membrane into the vacuole is controlled by two transporters. When zinc is abundant, it is taken up into the vacuole via these transporters, limiting its toxicity to the cell. When zinc is deficient, however, production of a third transporter, which mobilises removal of the sequestered zinc, is increased. A binding protein has been identified in yeast that regulates the genes responsible for encoding these zinc transporters. As in the examples above, this regulation is in response to zinc levels in the cell. Zn^{2+} is believed to be stored in the vacuoles of many plants and fungi and has also been detected in membrane-bound organelles in mammalian cells.

Zinc is of course not the only metal that is stored in this way. Other metals such as Mg^{2+}, Fe^{2+} and Ca^{2+} are also stored in the vacuoles of plants and fungi. We previously met the example of vanadium which has been found to accumulate in high concentrations in sea squirts (Box 4.7). The vanadium in one such species is stored within a highly acidic vacuole within the blood cell (vanadocyte) of the sea squirt. The vanadium is believed to enter the blood cell as vanadium(V), is then reduced to vanadium(IV) by reduced nicotinamide adenosine dinucleotide phosphate (NADPH) before final reduction to vanadium(III) in the vacuole. The low pH in the vacuole is maintained by a H^+-ATPase pump. (In addition, several vanadium binding proteins (vanabin) have been identified within the cytoplasm of the vanadocytes in some species of sea squirt, binding the vanadium as vanadium(IV).) In Chapter 4 you met the example of Ca^{2+} which is stored in the sarcoplasmic reticulum of muscle cells. Ca^{2+} is also believed to be stored more generally within the endoplasmic reticulum and within mitochondria for use in signalling.

5.5   Combating toxic levels of metals in cells

In the previous sections we have looked at some of the ways that metals are stored within cells. It is perhaps not surprising that some of these mechanisms for metal storage also come into play in detoxifying cells of toxic, non-nutrient metals. We will now consider some of the other strategies developed by organisms to prevent metal toxicity.

■   Many cells *exclude* metal ions by preventing their entry into the cytosol. How might this be achieved?

☐   One method is the regulation of transcription and/or translation in the synthesis of membrane transporters as in the examples above.

A further example of *exclusion*, demonstrated by some yeasts, is the secretion of ligands, such as sulfide, to form insoluble extracellular complexes with metals such as cadmium, thus reducing the bioavailability of the metal ions.

Another strategy, *extrusion*, is to pump the metal ions out of the cytoplasm. For example, in *E. coli*, accumulation of excess levels of zinc activates transcription of a pump that removes excess zinc from the cell. We saw another example of this in Chapter 1 where some microorganisms dispose of ingested mercury by converting it into the volatile $Hg(CH_3)_2$ which escapes through the bacterial membrane into the environment.

A third strategy is *detoxification*. One method is to render the metal harmless by chelation, as in the example of Cd^{2+} with metallothionein discussed briefly in Section 5.3. Another is to convert the metal itself into a less toxic form. One of the best understood examples is again mercury detoxification, in this case by some species of bacteria, which convert Hg(II) into Hg(0).

■ The Hg^{2+} ion is highly toxic to most organisms; can you suggest a reason why this might be so?

☐ A soft metal ion, Hg^{2+} has an extremely high affinity for sulfur, present in thiols, cysteines and other amino acids.

In addition, mercury can be metabolised to produce organomercury compounds, such as monomethyl mercury CH_3Hg^+, which are even more toxic. Hg(0), on the other hand, has little affinity for the ligands present in cells.

The bacteria release binding proteins that collect these organomercury compounds from the environment around the organism, and deliver them to the transport proteins, which take the mercury compound into the cytoplasm. The mercury–carbon bond in the organomercury compound is particularly stable to hydrolysis; however, the bacteria contain an enzyme, organomercurial lyase, which catalyses the process, increasing the rate of hydrolysis by approximately one million times.

■ What ligands might you expect to be present at the active site of this enzyme?

☐ Given the affinity of Hg^{2+} for sulfur, it is not surprising that the active site is believed to consist largely of sulfur-containing cysteine residues.

After hydrolysis, the resulting mercury(II) thiolate ($Hg(SR)_2$) is reduced by a second enzyme, mercuric-ion reductase, which reduces the Hg(II) to Hg(0). This enzyme also contains several cysteine residues to coordinate the Hg^{2+}.

Don't forget that there are questions on the companion website which you can use to test your understanding of the material covered in this chapter.

6 Biomineralisation

When you think of minerals, you probably think of rocks not animals and plants. However the sedimentary rocks chalk and limestone, which consist mainly of calcium carbonate in the form of calcite, originate from marine organisms. There are many examples of the use of calcite and other minerals by living creatures. Table 6.1 gives some examples of biominerals and their functions.

Table 6.1 Examples of biominerals and their functions.

Material	Form	Organism	Function
calcium carbonate	calcite	molluscs	shell (outer layer)
		mammals	gravity receptor in inner ear
		trilobites	eye lens
	aragonite	molluscs	shell (inner layer)
		fish	gravity receptor
	vaterite	sea squirts	protective spicules
calcium phosphate	hydroxyapatite	mammals	teeth
		vertebrates	bone
silica	amorphous	diatoms	exoskeleton
		molluscs (limpets)	teeth
		nettles	tips of hairs
iron oxides/hydroxides	magnetite	bacteria	magnetic sensor
	goethite	molluscs (limpets)	teeth
	lepidocrocite	molluscs (chitons)	teeth
	ferrihydrite	animals and plants	iron storage (ferritin)

Note the wide range of forms and uses. Shells are used for protection, bones to support bodies, and teeth to cut and grind food. Calcite crystals in the ear help us to recognise whether we are upright. In trilobites (an extinct arthropod found extensively as fossils), mineral crystals were used as lenses. Magnetotactic bacteria (which are found in both sea and freshwater) use a magnetic mineral to navigate. And of course in Chapter 5, you met ferritin, which stores iron as the hydrated iron(III) oxide, ferrihydrite.

In this chapter, we will consider some of the principles behind the bioproduction of minerals and look at a few examples of the process.

6.1 Introduction – repair of bone

Bone is a good example of a **biomineral**. To introduce some of the concepts needed to understand biomineralisation, let us take an overview of how broken bones are repaired by the body. Or more specifically from a chemical point of view: how are calcium ions selectively deposited onto the area of the break?

As we saw in Section 2.1.6, bone is a complex organic–inorganic composite. This illustrates an important feature of biominerals: organic, biological molecules play a significant role in their formation. In some cases, these molecules remain as part of the structure. In others, they are removed after mineral formation.

The organic component of bone, collagen, is a protein which occurs as filaments made up of three polypeptide chains; it has the higher-order structure shown in Figure 6.1a. In bone, these filaments of collagen are composed of five zones, four that are approximately 64 nm long and one of approximately 25 nm, Figure 6.1b. The filaments are arranged so that adjacent filaments are transposed 64 nm along their axis with respect to each other and cross-linked. Thus zone 1 on filament 1 is cross-linked to zone 2 on filament 2, etc. This results in a space at the ends of the molecules where plate-like crystals of hydroxyapatite, $Ca_{10}(PO_4)_6(OH)_2$, can grow.

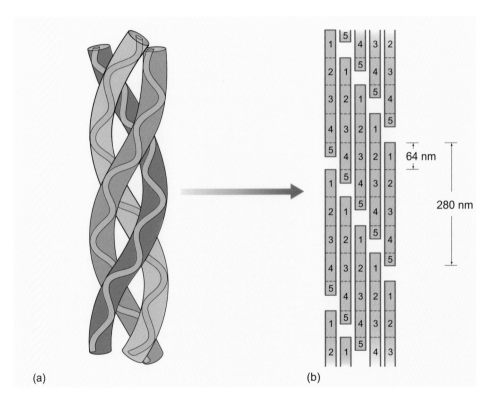

(a) (b)

Figure 6.1 (a) The higher-order structure of collagen showing three intertwined helical polypeptides and (b) alignment of collagen filaments showing zones.

Carboxylate-containing amino acid side chains are regularly spaced along the collagen fibres.

■ Do carboxylate-containing side chains act as hard or soft ligands?

☐ Carboxylate coordinates through oxygen; we expect these side chains to be hard, and therefore to coordinate to hard metal ions.

As you saw in Chapter 2, bone contains calcium ions, Ca^{2+} and, as Ca^{2+} is a hard metal ion, carboxylate groups should act as good ligands for it. This is part of the chemical process that selectively deposits calcium ions in bone. However, this is only part of the answer. Remarkably the spacing of carboxylate groups on collagen is such that the regular distance between neighbouring coordinated calcium ions is the same as that found in crystals of hydroxyapatite, $Ca_{10}(PO_4)_6(OH)_2$. As a result, small crystals of hydroxyapatite grow exclusively on the collagen fibres, resulting in the formation of bone (Figure 6.2).

This illustrates another general feature – the use of biological molecules as templates. In this case, the collagen not only provides a matrix for crystals to form on but also determines the shape and size of the crystals.

As well as illustrating some of the general features of biomineralisation, we can use the example of bone to pick out various topics we need to help us to understand the process. First we consider how crystals form and grow, and to do this you will need to become familiar with crystal structures and crystal surfaces. Then we look at what causes crystals to precipitate from solution and how, once this process has started, the crystals increase in size. Finally we consider how to regulate crystal growth. For example, why does bone consist of hydroxyapatite rather than any other form of hydrated calcium phosphate and why does the deposition of the mineral stop? The answer to this involves looking at promotion and inhibition of crystal growth.

We start by briefly reviewing crystal structure.

6.2 Crystal structure

In crystals, the atoms, ions or molecules are packed in regular arrays in three dimensions. The structure is usually represented by the repeating unit, the **unit cell**. Unit cells can be depicted in several ways. Perhaps the simplest to understand is a perspective drawing of the unit cell with the atoms represented by spheres, as shown for sodium chloride, NaCl, in Figure 6.3a. A packing diagram is a two-dimensional projection of the contents of the unit cell along one of the crystallographic axes. Figure 6.3b shows a packing diagram for NaCl along the z-axis. It is sometimes useful to view structures in terms of the coordination polyhedron around the metal. Figure 6.3c shows a depiction of the NaCl unit cell that illustrates the octahedral environment of the Na^+ ions. Yet other representations are used in some of the figures in this section.

The structure of sodium chloride is very simple. Biominerals such as hydroxyapatite have more complex structures. We shall now consider some of these.

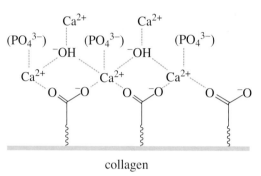

Figure 6.2 Schematic diagram of the biomineralisation process in bone. The regular spacing of the carboxylate amino acid side chains along the collagen polypeptide strand gives rise to regularly spaced calcium ions. The spacing is similar to that observed in single crystals of hydroxyapatite and so promotes the 'growth' of small crystals of hydroxyapatite on the collagen strand. Orange dotted lines indicate ionic bonding.

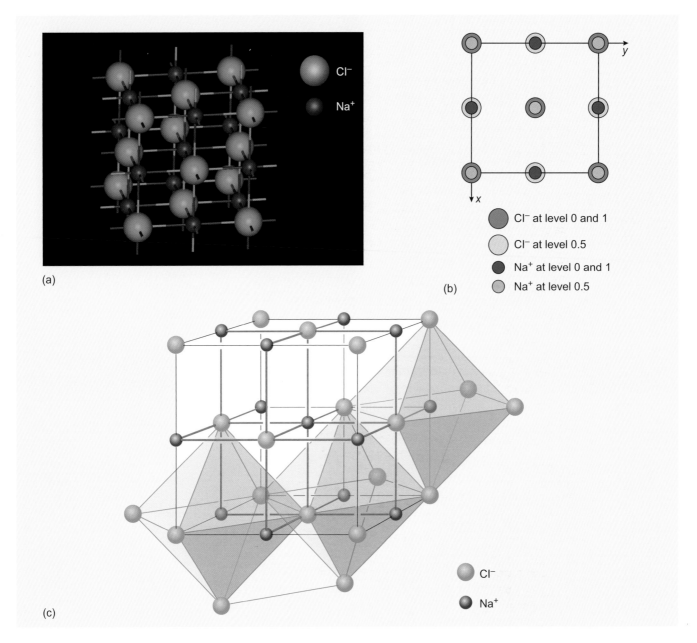

Figure 6.3 The unit cell of NaCl shown (a) as a perspective drawing, (b) as a packing diagram and (c) in terms of coordination polyhedra.

6.2.1 Calcium carbonate

Many seashells contain calcium carbonate, which is also the major constituent of limestone. Calcium carbonate can exist as several **polymorphs** (different crystal structures). Its most common and stable form at normal temperatures and pressures is **calcite**, shown in Figure 6.4a.

■ What shape are carbonate ions, CO_3^{2-}?

□ The ions are trigonal planar, that is, all the atoms are in the same plane and the three C–O distances are equal.

Carbonate ions occupy planes along the z-direction, here shown as horizontal, in between planes of Ca^{2+} ions. Within these planes, the carbonate ions are all oriented in the same direction. Carbonate ions in successive planes are oriented at 180° to each other.

Many marine organisms build their skeletons out of another polymorph, **aragonite**. This differs from calcite in that carbonate ions between the planes of Ca^{2+} ions are oriented differently (Figure 6.4b).

Finally there are a few examples of the biological use of a third polymorph, **vaterite**. In this, the plane of the carbonate anions is perpendicular to the Ca^{2+} ion planes (Figure 6.4c).

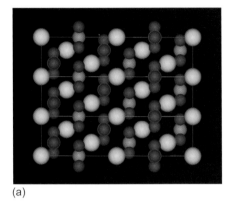
(a)
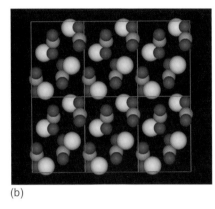
(b)
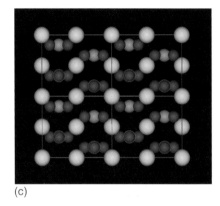
(c)

Figure 6.4 Crystal structures of the polymorphs of calcium carbonate: (a) calcite, (b) aragonite and (c) vaterite. The pale blue spheres represent calcium, red oxygen, grey carbon. In all these diagrams the z-axis is horizontal to show the Ca^{2+} ions lying in the xy plane as vertical rows. The cells are viewed along the x- or y-axes; each ion represents a row of ions along this direction. (Based on data from (b) Pkroy et al. (2007); (c) Meyer (1969).)

6.2.2 Hydroxyapatite

Hydroxyapatite forms the mineral component of bones and teeth. It is a phosphate hydroxide, formula $Ca_{10}(PO_4)_6(OH)_2$. Figure 6.5 shows the crystal structure of **hydroxyapatite**. The OH^- ions lie along the cell edges in channels surrounded by calcium ions.

In biological systems, hydroxyapatite often contains calcium **vacancies** and has a proportion of phosphate sites occupied by carbonate.

■ Why might substitution of phosphate by carbonate lead to calcium vacancies?

□ The charge on phosphate ions is −3 (PO_4^{3-}) and that on carbonate is −2 (CO_3^{2-}). Simply replacing phosphate by carbonate would therefore leave the crystal with a net positive charge. Neutrality can be restored by removing an appropriate number of Ca^{2+} ions.

In bone, there are often other variations on the composition with a range of cations and water substituting for some Ca^{2+} ions and a range of anions, water and vacancies substituting for some phosphate and some hydroxide ions. The basic structure is, however, that of hydroxyapatite.

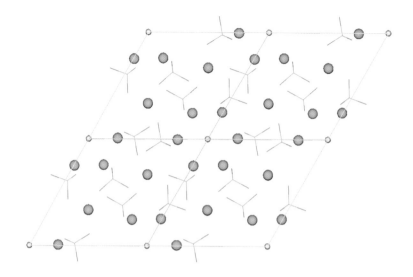

Figure 6.5 Crystal structure of hydroxyapatite. The blue-green spheres represent calcium, grey spheres are the H of hydroxide, red and orange structures are phosphate groups. (Based on data from Mostafa and Brown (2007).)

6.2.3 Iron oxides

A number of iron oxides and hydrated iron oxides are used by biological organisms. **Magnetite**, Fe_3O_4, is found in magnetotactic bacteria and in the teeth of certain molluscs known as chitons. The structure of magnetite is an example of a widely found crystal structure – the inverse spinel structure. Spinels have the general formula AB_2O_4 where A is a divalent cation and B a trivalent cation. The structure is based on a close-packed array of oxide ions with A ions occupying tetrahedral holes and B ions octahedral holes. In inverse spinels, the A ions occupy octahedral holes and the B ions are equally divided between octahedral and tetrahedral holes. For Fe_3O_4, shown in Figure 6.6, the A ions are Fe^{2+} and the B ions Fe^{3+}.

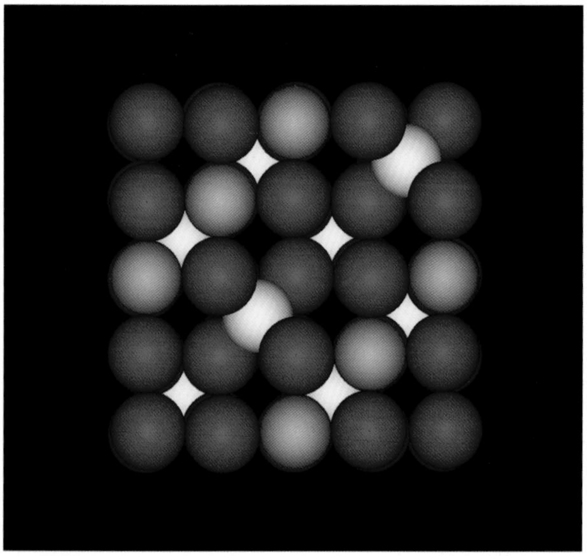

When the A and/or B ions have unpaired spins as in magnetite, this leads to interesting magnetic properties. The unpaired spins of the ions on the octahedral sites tend to align parallel to each other, as do those on the tetrahedral sites. The spins of the ions on the octahedral sites interact with those on the tetrahedral sites via the oxide ions and this leads to an antiparallel coupling. The coupling between the spins is sufficiently strong in magnetite that at room temperature almost all the spins on the octahedral sites are aligned in one direction and those on the tetrahedral sites in the opposite direction. The spins of the Fe^{3+} ions cancel out as half the ions are on octahedral sites and half on tetrahedral sites. The alignment of the Fe^{2+} ions, however, confers a net magnetic moment on the crystal. This property led to magnetite being used as an early compass (lodestone).

Figure 6.6 The structure of magnetite, Fe_3O_4.

The most stable oxide of Fe^{3+} is the mineral **hematite**, Fe_2O_3. However this only occurs as a biomineral in a hydrated form, $Fe_2O_3.nH_2O$. An important example of this is found in ferritin, which as you saw in Chapter 5, stores iron as the hydrated iron(III) oxide, ferrihydrite, $5Fe_2O_3.9H_2O$.

As well as magnetite, chiton teeth also contain the hydroxyl oxide **lepidocrocite**, γ-FeOOH. Another polymorph of this, **goethite**, α-FeOOH, is found in limpets' teeth. The two structures are shown in Figure 6.7.

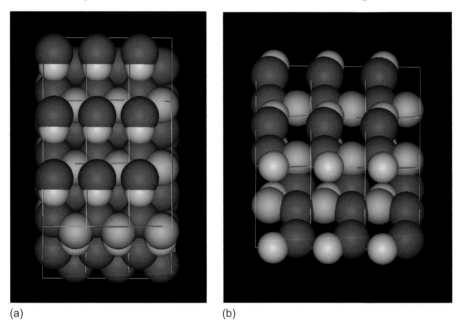

(a)   (b)

Figure 6.7 Crystal structures of (a) goethite and (b) lepidocrocite. (Based on data from (a) Hexiong et al. (2006); (b) Zhukhlistov (2001).)

These structures are very similar. They both contain planes of Fe^{3+} ions but the arrangement of OH^- ions in goethite is more regular.

6.3   Imperfect crystals

Most crystals are not exactly stoichiometric and perfectly ordered as we saw for hydroxyapatite. They deviate from perfection in several ways. **Point defects** occur when an ion is missing, is replaced by a different ion, moves to an off-lattice site or changes its oxidation state. You met examples of this in our discussion of bone in Section 6.1. The following are some of the most common forms of defect.

Intrinsic defects are integral to the crystal and do not change the composition. Two common intrinsic defects are **Schottky defects** and **Frenkel defects** (see Figure 6.8).

A Schottky defect consists of cation vacancies and compensating anion vacancies. Such defects are common in 1:1 compounds such as NaCl. The cation and anion vacancies are often randomly distributed through the crystal and do not necessarily occur in close proximity. In Frenkel defects, an ion moves from a lattice site to a vacant site, creating a vacancy and an

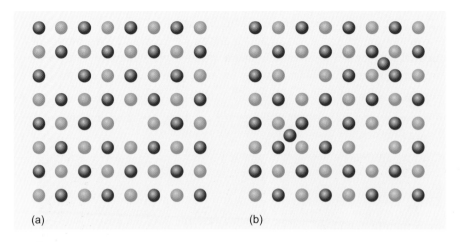

Figure 6.8   (a) A Schottky defect and (b) Frenkel defects.

interstitial ion. If an ion can change its valency, then it is also possible to form an intrinsic defect consisting of a vacancy plus an oxidation or reduction of an ion. For example, in FeO, Fe^{2+} vacancies are balanced by oxidation of some Fe^{2+} ions to Fe^{3+}.

Extrinsic point defects occur when an ion that is not part of the crystal formula substitutes on a lattice site or occupies an interstitial site. An important technological example, used in oxygen meters, is calcium-stabilised zirconia (ZrO_2) in which Ca^{2+} ions replace a small fraction of the Zr^{4+} ions. In bones and teeth, the substitution of CO_3^{2-} ions for a fraction of PO_4^{3-} ions in hydroxyapatite is an example of an extrinsic defect.

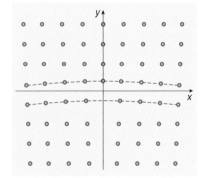

Figure 6.9   Sketch illustrating the formation of an edge dislocation. The atoms of the inserted plane are shown in red.

Crystals also show imperfections on a larger scale. Above nanometre size, crystals do not consist of one continuous lattice but are a collection of small regions, known as **crystal domains**, linked by grain boundaries. The ions within the grain boundaries are less ordered than in the domains. The disorder can mainly be described in terms of two types of dislocation. An **edge dislocation** occurs when an extra plane of atoms is inserted into part of a crystal (Figure 6.9). This causes distortion of the neighbouring planes.

The other important dislocation is the **screw dislocation** (Figure 6.10). For this a plane shifts relative to its neighbour for part of its length.

It was proposed by F. C. Frank at a seminal conference on crystal growth in 1949 that screw dislocations could provide a lower energy route for a crystal to grow (Frank, 1949). This is now widely accepted and you will meet this idea again in Section 6.5.6.

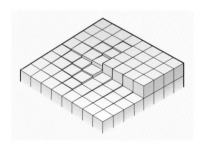

Figure 6.10   A screw dislocation.

6.4   Crystal surfaces

There is some evidence that biological systems contain organic structures which match a particular face of a mineral. In our example of the mending of bone in Section 6.1, the carboxylate ions in the collagen have spacings matching those of the Ca^{2+} ions found in a face of hydroxyapatite. Another example is nacre (mother of pearl) found on the inside of shells. This consists of sheets of aragonite sandwiched between sheets of protein–polysaccharide.

By matching a particular surface, an organism can control how a crystal grows and, in some cases, which polymorph is formed. In addition, the energy of interaction of surfaces with solvent plays a crucial role in crystal formation. Thus to understand how biominerals are formed, you need to know about surfaces.

The surface of a crystal is an exposed plane of atoms. In this section, we look at planes of atoms formed by slicing simple crystal structures and the notation used to describe these planes.

6.4.1 Surface of simple cubic structures

For simplicity we will focus initially on cubic structures formed from one type of atom only; many metals have cubic crystal structures and provide a suitable example.

Planes can be identified by three integers, h, k and l, which are related to the distances from the origin at which the plane cuts the x-, y- and z-axes, respectively. The three indices, h, k and l, are referred to as the **Miller indices** and a plane is described by the three indices enclosed in brackets (hkl).

Planes with Miller indices (100), (110) and (111) for a face-centred cubic (fcc) system are shown schematically in relation to the cubic unit cell in Figure 6.11.

A plane with indices (hkl) will intercept the x- and y-axes at $1/h$, $1/k$ and $1/l$ respectively. An index of 0 relates to a plane parallel to that axis.

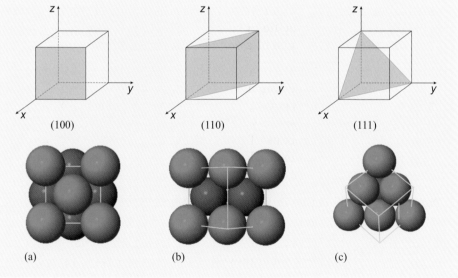

Figure 6.11 Planes in a face-centred cubic crystal with Miller indices (a) (100), (b) (110) and (c) (111) and their relationship to the cubic unit cell. The atoms shown in blue indicate the surface plane.

■ Consider the (100) plane shown in Figure 6.11a. Give the Miller indices of one of the other planes with the same two-dimensional array.

□ Any face of the cube will have the same two-dimensional array, for example, planes with Miller indices (010), (001), ($\bar{1}$00), (0$\bar{1}$0), (00$\bar{1}$)

where $\bar{1}$ represents -1. These are collectively represented using curly brackets $\{100\}$.

Note that the fcc(111) plane is a close-packed layer, that is, the available space is occupied as far as possible. In a close-packed two-dimensional array, each atom is surrounded by and touching six other atoms.

The three surfaces described have high thermodynamic stability *in vacuo* and are most likely to constitute the surfaces of small crystals. These surfaces have high atomic densities; indeed the (111) surface being close-packed must have the maximum density of any surface. Surfaces formed by planes with larger values of *h*, *k* and *l* have low atomic densities. For example, Figure 6.12 illustrates an fcc(332) surface shown in blue.

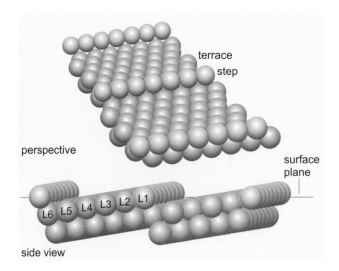

Figure 6.12   Perspective and side views of an fcc(332) surface, showing the appearance of terraces and steps. In each representation, the atoms in the surface layer are shown in blue. Successive atomic layers, from the surface layer downwards, are labelled on the side view.

Such surfaces can be viewed as a series of terraces of low-index structures ((100), (110) or (111)) separated by steps one atom in height. The (332) surface can be regarded as a set of stepped tilted (111) planes. You will see later in Section 6.5.2 that steps on the surface of crystals can be important sites for crystal growth.

6.4.2   Surfaces of an ionic compound, NaCl

So far we have only considered crystals composed of one sort of atom but, of course, the minerals we are interested in consist of several different ions. Still considering cubic structures, suppose we have a crystal that contains two ions such as sodium chloride. What do the low-index surfaces of this look like? The $\{100\}$ planes will lie parallel to a face of the unit cell. The surface will look similar to that of the face-centred cubic structure except that there are two sorts of ions in the layer (Figure 6.13a).

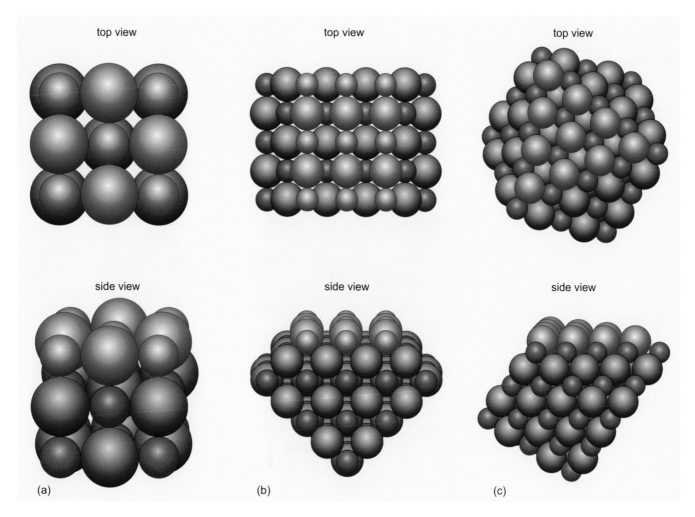

top view   top view   top view

side view   side view   side view

(a)   (b)   (c)

Figure 6.13 (a) (100) surface of NaCl, (b) (110) surface of NaCl and (c) (111) surface of NaCl. The green spheres represent Cl, mauve Na. The blue spheres show the surface layer.

Figure 6.13b shows the (110) surface of sodium chloride. This contains rows of the two types of ions, Na^+ and Cl^-, shown in blue on the surface. The next layer of the crystal contains both ions in the channels between these rows.

The last surface of NaCl that we will consider is the (111) surface (Figure 6.13c). This consists of a close-packed layer of one type of ion.

6.4.3 Surfaces of biominerals

Biominerals, as you have seen in Section 6.2, are more complex structures and the surfaces are correspondingly more complex. We shall consider one, comparatively simple, example.

As mentioned above, nacre (mother of pearl) consists of plates of aragonite sandwiched between layers of organic material. At the centre of the sheets of organic material, there are macromolecules resembling proteins found in silk fibres, called silk fibroin-like proteins. There is X-ray evidence to indicate that the (001) face of aragonite is parallel to the antiparallel β-pleated sheet of the silk fibroin-like proteins and that distances in the protein sheet match those between Ca^{2+} ions in the aragonite.

■   Which crystal axes will the (001) plane be parallel to?

□ $h = k = 0$, hence the plane will be parallel to the x- and y-axes.

We said earlier (Section 6.2.1) that in the structure of aragonite (Figure 6.4b), there are planes of Ca^{2+} ions parallel to the x- and y-axes. Part of this layer of Ca^{2+} ions is shown viewed along the z-axis in Figure 6.14a.

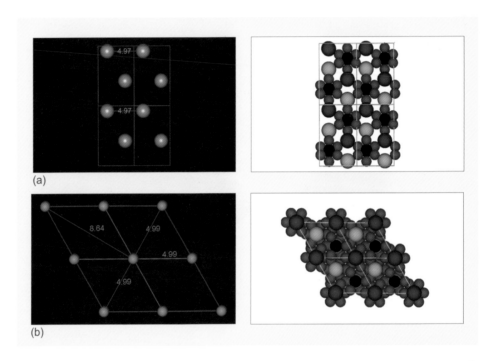

Figure 6.14   (001) surface of (a) aragonite and (b) calcite showing positions of Ca^{2+} ions. (Blue spheres Ca, red spheres O, black spheres C. Darker blue spheres indicate Ca on the (001) surface.) Note that the unit cell of calcite is not cubic and hence the view in (b) is not square.

The Ca^{2+} ions of the (001) surface of aragonite are 497 pm apart along the x-axis, the horizontal direction in Figure 6.14a. This is a good match to the repeat distance of the β-pleated sheet, 470 pm. Along the y-axis, the Ca^{2+} ions are 796 pm apart. This is larger than the spacing of oxygen atoms in the β-pleated sheet, 690 pm, but over longer distances, Ca^{2+} ions and oxygen atoms line up. For example, the spacing between a calcium ion and the eighth one along the row is approximately equal to the spacing between an oxygen atom in the β-pleated sheet and the ninth atom along (7 × 796 pm = 5572 pm, 8 × 690 pm = 5520 pm).

■   The spacing of the Ca^{2+} ions for calcite is 499 pm. Would calcite be a good fit for the β-pleated sheet?

□   Yes. The spacing is very similar to that along the x-axis in aragonite.

So, although the silk fibroin-like proteins do form a template for aragonite, simply matching the template is not the reason that nacre consists of aragonite rather than calcite. In Section 6.5, we will propose an additional factor.

6.5 Crystal formation

In this section, we consider what causes solids to precipitate out of solution and what determines the shape of crystals when they precipitate.

First we look at the driving force for precipitation.

In the presence of a solvent such as water, ionic solids dissolve until there is an equilibrium between the solid and the dissolved ions. The equilibrium constant describing this is the solubility product, K_{sp}.

■   Write an expression for the solubility product of calcium phosphate, $Ca_3(PO_4)_2$.

☐ $Ca_3(PO_4)_2(s) = 3Ca^{2+}(aq) + 2PO_4^{3-}(aq)$

As with other equilibrium constants if there is more than one mole of a substance in the equation, then we have to raise the concentration of that substance to a power in the expression of the equilibrium constant. Thus

$$K_{sp} = [Ca^{2+}(aq)]^3[PO_4^{3-}(aq)]^2$$

We have set this up for a solid dissolving but what we are actually interested in is the reverse – a solid precipitating out of solution. We can still use solubility products if we realise that they tell us what concentrations of ions have to be present for the solid to precipitate. Thus if the concentrations of calcium and phosphate ions in solution are such that their product exceeds the solubility constant, then we would predict that calcium phosphate would precipitate out.

6.5.1 Supersaturation

A solution in which the concentration(s) of dissolved substance(s) exceeds the values given by the solubility product is known as a **supersaturated solution**. You may be wondering how the concentrations can be higher than that given by K_{sp}. One way in which this can occur is that the solution is prepared at a temperature where the solubility product has a higher value and then the temperature adjusted. This is the basis of recrystallisation where the solvent is heated to dissolve the maximum amount of solid at a higher temperature and then cooled. At the lower temperature, the concentrations exceed those given by K_{sp} and crystals form. Another way of achieving supersaturation is to mix two solutions each containing one of the ions. For example lead nitrate and sodium chloride are both soluble, so that the concentrations of Pb^{2+} and Cl^- in separate solutions can easily exceed those given by K_{sp} for $PbCl_2$. When the solutions are mixed, the result is supersaturated with respect to $PbCl_2$ solid and so it precipitates out.

The first condition for a biomineral to precipitate is thus the formation of a supersaturated solution.

In biomineralisation, precipitation usually occurs within an enclosed space. There are several ways in which supersaturation can be achieved within this space.

- Either the cation or the anion can be selectively transported across a membrane or a boundary to increase the concentration within the enclosed space. For example, you saw in Chapter 5 that iron is transported into the ferritin protein.

- The ion can be supplied by transport of a soluble complex across membranes or a boundary into the enclosed space. The concentration of the free aqueous ion can then be increased to supersaturation levels by decomplexation. You will meet an example of this in magnetotactic bacteria (Section 6.7.3).

- Enzymes can release ions needed for the biomineral. Alkaline phosphatase, for example, releases HPO_4^{2-} at the sites of calcium phosphate mineralisation in cartilage.

- Concentrations can be increased by loss of water. Transpiration from the leaves of plants, which involves the loss of water, induces the deposition of silica.

- Precipitation can be encouraged by shifting other related equilibria. Phosphate and carbonate mineral solubility is strongly affected by changing the pH. This is because the phosphate and carbonate anions in solution are in equilibrium with their protonated forms and in the case of carbonate, with dissolved CO_2.

$$CO_2(aq) + H_2O(l) \rightleftharpoons (H_2CO_3(aq)) \rightleftharpoons H^+(aq) + HCO_3^-(aq)$$
$$\rightleftharpoons 2H^+(aq) + CO_3^{2-}(aq)$$

(6.1)

$$H_2PO_4^-(aq) \rightleftharpoons H^+(aq) + HPO_4^{2-}(aq) \rightleftharpoons 2H^+(aq) + PO_4^{3-}(aq)$$

(6.2)

The solubility of sparingly soluble salts is also affected by the presence of other ions in solution.

This last point is illustrated by the solubility of silver chloride, AgCl. Figure 6.15 compares the solubility of AgCl in aqueous solutions of magnesium sulfate with its solubility in pure water. Note that there is quite an increase in the solubility of AgCl with increasing concentration of $MgSO_4$.

The way in which other ions in solution affect solubility is explored in the next section.

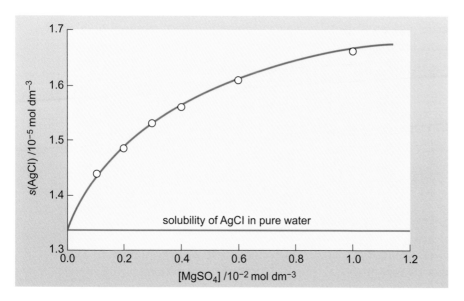

Figure 6.15 The solubility of AgCl, in aqueous solutions of MgSO$_4$ of various concentrations at 298.15 K.

6.5.2 Ions in solution

Silver chloride is 'sparingly soluble' in water. Nevertheless, when AgCl is shaken up with water, some of the solid does dissolve, forming a saturated solution: the equilibrium between the solid and its aqueous ions can be represented as:

$$AgCl(s) = Ag^+(aq) + Cl^-(aq) \tag{6.3}$$

The extent to which the salt dissolves is described by the solubility constant, K_{sp}. For AgCl,

$$K_{sp} = [Ag^+(aq)][Cl^-(aq)] \tag{6.4}$$

At 298.15 K, this has the value of 1.74×10^{-10} mol^2 dm^{-6}.

However, when silver chloride is dissolved in an aqueous solution of a different electrolyte – magnesium sulfate (MgSO$_4$), for example – the situation changes markedly. As Figure 6.15 shows, the solubility of AgCl is now different from that in pure water and dependent on the concentration of the added electrolyte: in this case, the effect is quite substantial.

■ What does this suggest about the value of K_{sp}?

☐ If the solubility of AgCl changes, then so too must the product of the concentrations of Ag$^+$(aq) and Cl$^-$(aq). Thus K_{sp} does not appear to hold constant.

Arguments based on thermodynamics show that the definition of solubility product in terms of concentrations is an approximation that really only applies if there is no interaction between the ions in solution. That is, when the solid dissolves, it forms isolated ions that interact only with solvent. A solution in which the solute particles (ions or molecules) are non-interacting is known as an **ideal solution**. Beyond extremely dilute solutions, the assumption of ideality does not hold. In solutions of ionic salts, for example, the ions interact via electrostatic interaction and may also form clusters of ions.

Now, we can define an equilibrium constant for the dissolution of solids that is only dependent on temperature and not on the concentration of ions in the solution. This equilibrium constant is the **standard solubility product**, $K_{sp}^{\ominus}$. This is defined in terms of a quantity known as **activity**, a, rather than concentration.

Since the activity of the solid AgCl is unity, $K_{sp}^{\ominus}$ for silver chloride becomes

$$K_{sp}^{\ominus} = a(\text{Ag}^+)a(\text{Cl}^-)$$

(6.5)

The activity, a, is given by $a = \gamma c / c^{\ominus}$, where c is the concentration, $c^{\ominus}$ is a standard concentration of 1 mol dm^{-3} and the factor γ is the **activity coefficient**. γ describes the deviation from ideality, that is, it allows for the interaction between solute particles.

There is, however, a problem with obtaining activity coefficients for individual ions. A solution of AgCl (or any other electrolyte for that matter) contains both cations and anions: there is no way of 'disentangling' the product $\gamma_+\gamma_-$ experimentally, and of assigning one part to the cations and another to the anions. In other words, individual ionic activity coefficients cannot be measured.

The way around this difficulty is to define the activity of an ion only in the presence of 'counter-ions' (that is, ions of opposite charge). In practice, this is achieved by introducing a **mean ionic activity coefficient**, $\gamma_{\pm}$. For a 1:1 electrolyte, such as AgCl or MgSO$_4$, it is defined as:

$$\gamma_{\pm} = (\gamma_+\gamma_-)^{1/2}$$

(6.6)

Experimental values of mean ionic activity coefficients for various (relatively soluble) electrolytes are given in Table 6.2: evidently, as you might expect, the value of $\gamma_{\pm}$ for a given electrolyte depends strongly on the concentration of the solution. Although the detailed trends in Table 6.2 are far from simple, broadly speaking the more concentrated the solution, the greater the deviations from ideality – that is, the more different from unity is $\gamma_{\pm}$. And the effect appears to be more pronounced if one (or both) of the ions is highly charged.

The clue to what is happening lies in the fact that ions are charged: oppositely charged ions attract one another. Although the solution is neutral overall, this suggests that cations and anions are unlikely to be distributed completely at random, and therefore uniformly, throughout the solution. Rather, we would expect to find a predominance of anions in the vicinity of cations, and vice versa. We shall explore this in the next section.

Table 6.2 Mean ionic activity coefficients, $\gamma_{\pm}$, for various electrolytes as a function of their concentration, c (in aqueous solution at 298.15 K).

c/mol dm^{-3}	$\gamma_{\pm}$ (HCl)	$\gamma_{\pm}$ (NaOH)	$\gamma_{\pm}$ (CaCl$_2$)	$\gamma_{\pm}$ (H$_2$SO$_4$)	$\gamma_{\pm}$ (ZnSO$_4$)	$\gamma_{\pm}$ (Al$_2$(SO$_4$)$_3$)
0.001	0.996	–	0.888	0.830	0.734	–
0.01	0.904	–	0.732	0.544	0.387	–
0.10	0.796	0.766	0.524	0.265	0.147	(0.035)
0.50	0.758	0.693	0.510	0.155	0.063	0.014
1.00	0.809	0.679	0.725	0.130	0.043	0.018
2.00	1.010	0.700	1.554	0.124	0.035	–

6.5.3 The Debye–Hückel theory

Suppose now that Figure 6.16a is a snapshot of the ionic distribution at a given instant in time. Ions are constantly on the move, so different snapshots would show different distributions. Nevertheless, scrutiny of a sufficient number of pictures would allow recognition of a certain time-average distribution, in which any given ion would be surrounded by a 'spherical haze' of opposite charge – the so-called ionic atmosphere of the ion (Figure 6.16b).

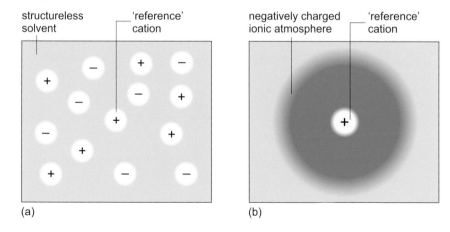

(a)   (b)

Figure 6.16 (a) Representation of the distribution of ions in solution, (b) an ion surrounded by an ionic atmosphere.

This simple picture of an ionic solution underlies the interpretation of ionic activity coefficients, according to a molecular model first put forward by P. Debye and E. Hückel in 1923. Its genius was its simplicity as the theory argues that the deviations from ideality (contained in $\gamma_{\pm}$) are due solely to electrostatic interactions among the ions present – interactions that lead to the sort of average distribution indicated in Figure 6.16b. The **Debye–Hückel theory** argues that the level of electrostatic interactions is a function not only of the concentrations of all the ions in solution but also of the charges on those ions. This certainly ties in with our comment about the experimental values of $\gamma_\pm$ in Table 6.2.

The final result of that analysis, the Debye-Hückel theory, states that

$$\log \gamma_\pm = -A|z_+z_-|I^{1/2} \tag{6.7}$$

where z_+ and z_- are just the charges on the positive and negative ions, respectively; the modulus sign $|\ |$ means the magnitude of the product z_+z_-, without its sign (thus for AgCl, for example, $z_+ = +1$, $z_- = -1$ so $z_+z_- = -1$ but $|z_+z_-| = 1$); A is a constant, the value of which depends on certain physical properties of the solvent (notably its density and dielectric constant) and the temperature; and finally the term I is known as the ionic strength. I is defined as:

$$I = \frac{1}{2}\sum_i \frac{c_i z_i^2}{c^\ominus} \tag{6.8}$$

where c_i is the concentration of ion i of charge z_i, and $c^\ominus$ is the standard concentration (1 mol dm^{-3}).

For a solution of a single 1:1 electrolyte of univalent ions – such as AgCl or NaOH – the ionic strength is just the concentration of the electrolyte (or strictly, the numerical magnitude of that concentration).

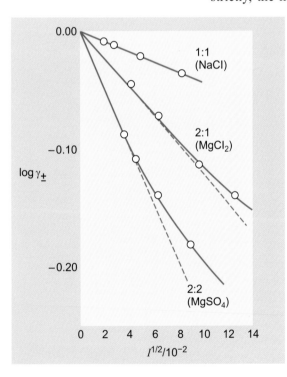

But this is not always so; in general, the link between concentration and ionic strength depends on the valence types of the ions – even for a solution containing a single electrolyte. More importantly, in a solution containing several electrolytes, each ionic species makes its own contribution to the overall ionic strength.

Thus, to determine the ionic strength of a solution, all the individual values of $c_i z_i^2$ must be added together, one term for each species of ion present in the solution.

When another ionic solid is added to a solution of a sparingly soluble salt, then the ions in solution will interact with the added ions as well as the ions from the salt itself. This extra interaction will alter the value of the activity coefficient $\gamma_\pm$.

For water at 298.15 K, we can simplify Equation 6.7. For these conditions, $A = 0.51$, and Equation 6.7 becomes

$$\log \gamma_\pm = -0.51|z_+z_-|I^{1/2} \tag{6.9}$$

Although the solubility of ionic salts is always affected by the ionic strength, the actual relationship in Equation 6.9 only holds for very dilute solutions and is hence known as the **Debye–Hückel limiting law**. Figure 6.17 shows the predicted dependence of log $\gamma_\pm$ on I for different electrolytes, compared with that found experimentally. A divergence between theory

Figure 6.17   The observed (full line) dependence of the mean ionic activity coefficient on ionic strength for various valence types compared with the predictions (dashed line) of the Debye–Hückel limiting law.

and experiment is already apparent for the higher valence types at an ionic strength of around 0.2 (as $I^{1/2} = 0.04$).

■ What is the concentration of $CaCO_3$ in a solution containing this salt with ionic strength 0.2?

☐ If c is the concentration of the salt and hence of both Ca^{2+} ions and $CO_3{}^{2-}$ ions, then,

$$I = \frac{1}{2}\left[\frac{(+2)^2 c}{c^{\ominus}} + \frac{(-2)^2 c}{c^{\ominus}}\right] = 4c$$

Hence $4c = 0.2$ and $c = 0.05$ mol dm^{-3}.

At higher values of the ionic strength, deviations are apparent even for 1:1 electrolytes.

6.5.4   Solubility in biological fluids

Now it is possible to see how in biological systems, control of precipitation could be achieved by increasing or decreasing the concentration of non-precipitating ions such as Na^+, K^+ and Cl^-.

The ionic strength of blood plasma is 0.15 due to the presence of ions of soluble salts such as NaCl. Let us see what difference an ionic strength of 0.15 makes to the concentration of Ca^{2+} ions in a saturated solution of calcite.

The solubility product of calcite, $K^{\ominus}_{sp} = 3.31 \times 10^{-9} = 10^{-8.48}$. In pure water, the only ions are Ca^{2+} and $CO_3{}^{2-}$. The solubility product is very small and so the ionic strength is very low. Hence the concentration is a good approximation to the activity.

Thus $[Ca^{2+}][CO_3{}^{2-}] = [Ca^{2+}]^2 = 3.31 \times 10^{-9}$ mol^2 dm^{-6}

and so $[Ca^{2+}] = (3.31 \times 10^{-9})^{0.5} = 5.75 \times 10^{-5}$ mol dm^{-3}

Now if the ionic strength is 0.15, then from Equation 6.9,

$$\log \gamma_{\pm} = -0.51 \times 4 \times 0.15^{0.5} = -0.51 \times 4 \times 0.387 = -0.79$$

and so, $\gamma_{\pm} = 10^{-0.79} = 0.16$.

In terms of activity,

$$K^{\ominus}_{sp} = a(Ca^{2+})a(CO_3{}^{2-}) = \frac{\gamma_{\pm}^2[Ca^{2+}]\,[CO_3{}^{2-}]}{c^{\ominus}} = \frac{\gamma_{\pm}^2[Ca^{2+}]^2}{c^{\ominus\,2}}$$

Thus

$$3.31 \times 10^{-9} = (0.16)^2 \frac{[Ca^{2+}]^2}{c^{\ominus\,2}}$$

and so $[Ca^{2+}] = 3.60 \times 10^{-4}$ mol dm^{-3}.

This is over 6 times the concentration in pure water. Thus calcite will be more soluble in blood plasma than in pure water and a greater concentration will be needed before calcite will be precipitated.

6.5.5 Nucleation

Although thermodynamically a supersaturated solution must produce a precipitate, in practice this can be a slow process. There has to be a mechanism by which crystals can form. The initiating process is **nucleation**. This can occur spontaneously or through the presence of a nucleation site in the solution.

Spontaneous nucleation arises as follows. Small clusters of ions are continually forming and falling apart in solution. It is only when they reach a critical size that precipitation occurs. This size is determined by a balance between Gibbs free energy terms.

The bonding between ions in the cluster makes the process of clustering thermodynamically favourable. This bonding increases with the size of the cluster and the energy gained is proportional to the volume of the cluster.

Opposing this is the Gibbs free energy required to form a solid–liquid interface. This is positive and depends on the surface area.

Both the volume and surface area increase with cluster size but the volume increases more rapidly as we can illustrate by considering a spherical cluster.

■   What are the volume and surface area of a sphere of radius, r?

☐   The volume is given by $\frac{4}{3}\pi r^3$ and the surface area by $4\pi r^2$.

The Gibbs free energy due to bonding thus decreases as r^3 and that due to the formation of a solid–liquid surface increases as r^2.

■   What is the ratio of the surface area to the volume for $r = 2$ and for $r = 10$?

☐   The ratio is given by $4\pi r^2/(\frac{4}{3})\pi r^3 = 3/r$. So, for $r = 2$ the ratio is 1.5 and for $r = 10$ it is 0.3.

Thus, for small values of r, the surface area will be large compared to the volume and the solid–liquid interface term dominates. As the size increases, the Gibbs free energy reaches a maximum and then decreases as the bonding term starts to dominate. The radius for which the maximum occurs is the **critical radius**, r_c.

So, what controls nucleation?

The first factor controlling nucleation is the concentration of ions in the supersaturated solution. The more these exceed the concentrations needed to satisfy K_{sp}, the more likely is the formation of larger clusters and hence the rate of precipitation. Both slow and fast rates of precipitation can be advantageous. To form large crystals, it is necessary to grow the crystals slowly and so it would be important to have concentrations that are not too far above those in a saturated solution. On the other hand, small crystals or rapid precipitation may be required and a highly supersaturated solution is then more appropriate. We shall see below that the degree of supersaturation can affect not only the rate of precipitation but also the nature of the polymorph that precipitates out.

The second is the value of the solid–liquid interface energy per unit area. Specks of dust can provide a surface to which clusters can attach and can reduce this energy significantly, thus allowing smaller clusters to be more stable. For example, foreign bodies in oyster shells form a centre of nucleation for pearls. Small particles are particularly effective if they are similar in chemical composition and stereochemistry to the mineral. Such surfaces can also be provided by the containing vessel or by seed crystals or templates. Biominerals often nucleate on templates formed from organic macromolecules such as proteins. You saw in Section 6.4 that nacre used silk fibroin-like proteins.

An interesting and at first sight puzzling observation is that at very high degrees of supersaturation, the most soluble form of a salt crystallises out first. This observation is known as the **Ostwald–Lussac law** and is the result of an interplay of kinetic and thermodynamic factors. The more soluble a form of the salt, the less favourable its lattice energy but the more favourable its solid–liquid interface energy. At the initial stage of nucleation, clusters of the most soluble form of the salt are produced preferentially due to the lower interface energy. In highly supersaturated solutions, these precipitate out as crystals of the most soluble polymorph or even as amorphous solids since these are generally more soluble than crystalline forms of a solid. In weakly supersaturated solutions, crystallisation is slower and the product is the thermodynamically favourable least soluble polymorph.

■ The $K_{sp}^{\ominus}$ values of various forms of calcium carbonate are: calcite 3.31×10^{-9}, aragonite 4.57×10^{-9} and vaterite 1.23×10^{-8}. Which of these would you expect to precipitate out from (a) weakly supersaturated solutions and (b) highly supersaturated solutions?

□ (a) In weakly supersaturated solutions, the least soluble crystallises out. This is the thermodynamically most stable polymorph, calcite. (b) The most soluble polymorph precipitates from highly supersaturated solutions. This is vaterite or possibly amorphous calcium carbonate.

6.5.6 Crystal growth

Crystal growth is the addition of extra layers of ions to the cluster formed in the nucleation process. Table 6.3 shows the linear rate of growth of several ionic crystals from solution. The quantity S in the table is the degree of supersaturation expressed as c/c_0 where c is the concentration in kilograms of solid per kilogram of water and c_0 is the concentration in these units of the saturated solution at the temperature given. v_L is the mean linear growth velocity.

■ How does the linear growth velocity vary with degree of supersaturation and temperature? Is this what you would expect?

□   The velocity increases with degree of supersaturation and with temperature. This is what you would expect for a rate process.

Table 6.3   Linear growth rate of ionic crystals.

Solid	$T/°C$	S	$v_L/m\ s^{-1}$
NaCl	50	1.002	2.5×10^{-8}
	50	1.003	6.5×10^{-8}
	70	1.002	9.0×10^{-8}
	70	1.003	1.5×10^{-7}
$NH_4H_2PO_4$	30	1.02	3.0×10^{-8}
	30	1.05	1.1×10^{-7}
	40	1.02	7.0×10^{-8}

■   From the data in Table 6.3, how long would it take to increase the size of a crystal of NaCl in a solution with $S = 1.003$ at 70 °C by 1 mm?

□   The mean linear velocity is $1.5 \times 10^{-7}\ m\ s^{-1}$. An increase in linear size of 1 mm would take $1 \times 10^{-3}\ m/1.5 \times 10^{-7}\ m\ s^{-1} = 6.7 \times 10^3$ s. This is 1.9 hours.

However, the growth rate is not simply dependent on temperature and degree of supersaturation as we will see below.

First, we have to remember that ions in solution are hydrated, whereas not all crystals contain water molecules. So we have to consider diffusion of the hydrated ions to the surface of the crystal. Then the ions have to diffuse along the surface and possibly within the crystal. Generally there will be loss of water of hydration and finally incorporation of the bare ion into the crystal. So the rate of growth will also depend on diffusion rates both through the solution and through the crystal. The rate of loss of coordinated water will also contribute.

Second, laying down more ions on a perfect crystal surface is slower than adding ions to the surface at sites of imperfection. These may be steps or kinks of the sort you saw in the low atomic density (high h, k and l) planes (Section 6.4) or they could be local defects such as impurities deposited in the crystal nucleus. For low levels of supersaturation, it has been proposed that crystals grow on dislocations in the crystal, particularly screw dislocations.

In addition, other ions or molecules in the solution can also either aid (as **accelerators**) or inhibit (as **inhibitors**) growth. This third factor is particularly important for living systems as most biological fluids are saturated with respect to biominerals. If growth were purely controlled by the first two factors, then structures such as teeth and shells could carry on growing and become harmful to the organism. Thus biological systems frequently regulate growth by using inhibitors and accelerators.

6.5.7   Growth of phosphates

A well-studied example of the effects of inhibitors and accelerators is the crystallisation of calcium phosphates. Hydroxyapatite is just one possible calcium phosphate mineral. The formulae and solubilities of other minerals are given in Table 6.4.

Table 6.4   Calcium phosphate minerals and their solubilities.

Mineral	Formula	$K_{sp}^{\ominus}$	$[Ca^{2+}]/\text{mol dm}^{-3}$
brushite	$CaHPO_4.2H_2O$	1.87×10^{-7}	5.7×10^{-3}
octacalcium phosphate	$Ca_8(HPO_4)_2(PO_4)_4.5H_2O$	2.6×10^{-99}	1.35×10^{-7}
tricalcium phosphate	$Ca_3(PO_4)_2$	2.8×10^{-29}	2.3×10^{-6}
hydroxyapatite	$Ca_{10}(PO_4)_6(OH)_2$	5.5×10^{-118}	4.3×10^{-7}
monetite	$CaHPO_4$	9.2×10^{-7}	9.6×10^{-4}
fluorapatite	$Ca_{10}(PO_4)_6F_2$	1.0×10^{-118}	3.95×10^{-7}

The solubility products are very small numbers. However remember that each formula in Table 6.4 contains a large number of ions. The solubility product of hydroxyapatite, for example, will be given by $[Ca^{2+}(aq)]^{10}[PO_4^{3-}(aq)]^6[OH^-(aq)]^2$. The final column in Table 6.4 gives the concentrations of Ca^{2+} ions assuming only ions from the dissolved mineral are present. The concentration of Ca^{2+} ions in saliva is about 10^{-6} mol dm^{-3}.

There is evidence that the growth of calcium phosphate biominerals starts from nucleation of amorphous calcium phosphate. As amorphous forms are often more soluble than crystalline forms of a solid, in highly supersaturated solutions, according to the Ostwald–Lussac law, these forms will crystallise first. In the laboratory, at high concentrations of Ca^{2+} and PO_4^{3-} ions, amorphous calcium phosphate of formula $Ca_3(PO_4)_{1.87}(HPO_4)_{0.2}.nH_2O$ precipitates out. On standing, the amorphous phase redissolves and octacalcium phosphate crystals grow. Octacalcium phosphate crystallises as plates. Its crystal structure consists of layers of Ca^{2+} and PO_4^{3-} resembling those of hydroxyapatite interleaved with layers in which these two ions are more widely separated. It has been proposed that octacalcium phosphate is an intermediate in the formation of hydroxyapatite in bone and even acts as a template for the plate-like form of the crystals. The ^{31}P NMR spectrum of bone is a broad peak covering the chemical shift of octacalcium phosphate as well as that of hydroxyapatite (Duer, 2008).

The proposed overall mineralisation pathway for phosphate biominerals is:

amorphous calcium phosphate → brushite → octacalcium phosphate → tricalcium phosphate → hydroxyapatite

As mentioned above, most biological fluids are supersaturated with respect to calcium phosphate minerals. However there are several inhibitors that block the route from amorphous calcium phosphate to the crystalline forms.

Inhibitors of hydroxyapatite formation include Mg^{2+}, CO_3^{2-}, ATP and ADP. Formation of phosphate minerals in bones and teeth is controlled by release of enzymes that temporarily remove inhibitors allowing the desired biomineral to grow in a controlled manner.

6.5.8 Shapes of crystals

A further aspect of crystal growth is the shape that the crystal adopts.

As a crystal grows, it will grow faster in the direction perpendicular to one or more faces than in directions perpendicular to others. The relative rates of growth along the different faces determine the shape (or habit) of a crystal.

Different faces of a crystal have different surface energies, so that the energy barrier to growth is higher on some surfaces than others. Surfaces with higher energy barriers will grow more slowly.

In addition kinks and dislocations are more likely to occur on some faces, so that these faces will grow at the expense of the others. Impurities can alter the habit of a crystal by, for example, selectively attaching to one face and thereby stopping or accelerating growth of that face. For example, the hexacyanoferrate(II) ion, $[Fe(CN)_6]^{4-}$, can cause sodium chloride to crystallise as branching tree-like (dendritic) crystals rather than cubic crystals and at slightly higher concentrations suppresses the growth of NaCl crystals altogether.

Having looked at the factors concerned with the nucleation and growth of biominerals, we shall now turn to some specific examples. Our first example is teeth.

6.6 Human teeth

Figure 6.18 shows a sketch of the structure of a tooth. The outer layer above the gum is enamel. Beneath that is dentine. Between the dentine and the jawbone, and connecting the tooth to the jawbone, are cementum (a composite bone–dentine material) and the peridontal ligament.

Both dentine and enamel contain forms of the mineral hydroxyapatite. Dentine is similar to bone in that it contains collagen and the hydroxyapatite crystals are of similar size – of the order of tens of nanometres. Enamel contains much less organic material and the mineral crystals are about ten times larger in all directions. Dentine forms in a similar way to bone; as in bone, collagen provides a template on which the crystals of hydroxyapatite form.

Two types of organic macromolecules are involved in the growth of enamel: hydrophobic macromolecules called **amelogenins** and acidic macromolecules known as **enamelins**. Amelogenins are proteins with a relatively large number of neutral amino acid side chains – proline, glutamine, leucine and histidine. They form spheres with diameters of the order of nanometres that bind to specific crystal planes blocking these from further growth. This causes the hydroxyapatite crystals to grow preferentially along the z-axis forming long, thin crystals. These crystals are interwoven like cloth to give them extra stability. As the crystals develop, the amelogenins are removed, leaving a

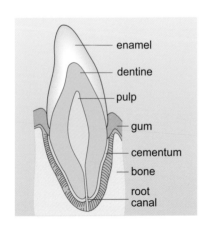

Figure 6.18 Section through a tooth showing enamel, dentine, cementum and pulp.

material with a high mineral content. After removal of the amelogenins, the enamel crystals grow rapidly along the x- and y-axes.

Enamelins are proteins attached to polysaccharides with sulfate and carboxylate groups.

■ The protein component of enamelins contains a high proportion of glutamate and aspartate residues. What type of side chain do these residues have?

☐ The side chains are charged.

Although the exact function of enamelins is still a subject of research, they are believed to play a key role in the growth of hydroxapatite crystals that form enamel. They form a sheath around the growing crystals which is mostly removed when the tooth is mature to give enamel a composition of 95% hydroxyapatite.

■ Saliva is supersaturated with respect to hydroxyapatite and yet our teeth (unlike those of rabbits) are not continually growing. Suggest how the growth may be prevented.

☐ Biological fluids contain inhibitors and a reasonable suggestion is that when the teeth are mature, the enzymes counteracting the inhibitors are no longer released and the inhibitors are free to prevent growth.

In enamel, growth is inhibited by phosphoproteins which bind to the surface.

Although our teeth stop growing, they are subject to erosion. They can be ground down by hard substances such as bits of stone in stoneground flour. You should note also that the formula of hydroxyapatite contains OH^- ions.

■ What does this suggest might affect the solubility of this mineral?

☐ pH will affect the solubility.

■ Would you expect hydroxyapatite to be more or less soluble in acid solution?

☐ When hydroxyapatite dissolves, OH^- ions are produced. If the concentration of H^+ ions in the acid solution is sufficiently high, then these will react with the hydroxide ions to form water. More hydroxyapatite will dissolve to restore equilibrium. So the solubility of hydroxyapatite will increase with decreasing pH.

This has an unfortunate consequence for our teeth in that acid foods and drinks (such as fizzy drinks) will lower the pH so that tooth enamel starts to dissolve. (This is sometimes referred to as acid erosion.)

In recent years (2009) a number of brands of toothpaste have made claims regarding the ability to 'restore' enamel by adding microparticles of hydroxyapatite or other sources of calcium and phosphate to the toothpaste. A more substantiated approach is the use of fluoride in toothpaste or drinking water. In the presence of fluoride ions, the dissolved hydroxyapatite in saliva precipitates as fluoroapatite, in which the OH^- ions in the calcium channels

(Figure 6.5) are replaced by F^- ions. Fluoroapatite is less soluble than hydroxyapatite (see Table 6.4) and forms a tenacious coating that is effective in preventing tooth decay. Interestingly, the enamel of shark's teeth naturally contains a much greater percentage of fluoroapatite than that of human teeth.

In Section 6.7.2, we will consider teeth that contain iron oxide, but first we revisit the storage of iron in the body.

6.7   Iron oxides – ferrihydrite, limpet teeth and bacterial magnets

As you saw in Table 6.1, there are a number of iron oxide/hydroxide biominerals. One of these, ferrihydrite, the mineral found in ferritin, is thought to play a crucial role in the formation of iron oxide biominerals as it can be converted into the other minerals. Ferrihydrite has the formula $Fe_2O_3.nH_2O$, but can be considered an oxide/hydroxide of iron. As well as occurring as a biomineral, it is readily produced in a test tube as a brown, gelatinous precipitate by adding sodium hydroxide solution to a solution of an Fe(III) salt. It is more soluble than the other iron oxide biominerals and is amorphous.

■   What does its relative solubility suggest according to the Ostwald–Lussac law?

☐   In highly supersaturated solutions, this will be preferentially precipitated.

Precipitated ferrihydrite can be transformed into hematite (Fe_2O_3) and the biominerals goethite, lepidocrocite and magnetite.

■   Where have you met before a series of related minerals for which an amorphous form initially crystallises and can be successively transformed into the other minerals?

☐   The calcium phosphate minerals (Section 6.5.7).

First we consider ferrihydrite in ferritin. Then we move on to other iron-containing minerals, some of which are produced via ferrihydrite.

6.7.1   Ferritin revisited

In ferritin, ferrihydrite is contained in polypeptide cages of 8 nm internal diameter. As you saw in Chapter 5, the polypeptide cavity housing the stored iron has several channels about 300 pm wide that go through the coating of the shell. Some of these are hydrophobic and some hydrophilic. Fe^{2+} is transported into the cavity via the hydrophilic channels, which are lined with aspartate and glutamate residues. Fe^{2+} binds to several amino acid residues (glutamic acid, tyrosine, histidine and glutamine) at the ferroxidase centre in the polypeptide shell. It is then oxidised by O_2 to Fe^{3+}. The Fe^{3+} ions are attracted to the surface of the cavity where there are three glutamate residues close together and another three not far away on an adjacent strand.

■   Why might Fe^{3+} ions be attracted to this part of the cavity?

- ☐ Glutamate residues have COO⁻ side chains and so a cluster of three forms a highly negatively charged environment to which the Fe^{3+} ions would be attracted.

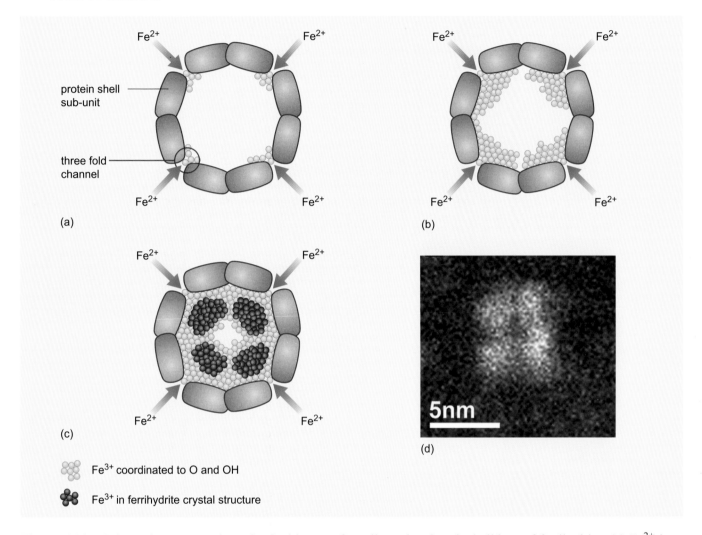

Figure 6.19   Schematic cross-section of a ferritin core from liver showing the build-up of ferrihydrite: (a) Fe^{2+} ions deposit near the threefold symmetry entry channels. These are then oxidised to Fe^{3+}; (b) more Fe^{2+} ions enter and are oxidised; and (c) Fe^{3+} diffuses inwards and forms ferrihydrite. (d) Electron microscope image of the iron core (white areas) showing the fourfold symmetry (Pan et al., 2009).

EXAFS studies of haemosiderin extracted from cells, which can be regarded as a degraded form of ferritin, indicated that the iron is present as an amorphous iron oxide/hydroxide. Electron microscopy of haemosiderin and ferritin within cells, however, indicates that the iron is stored as polycrystalline ferrihydrite. A recent electron microscope study (Pan et al., 2009) indicates that the iron-containing core has a regular structure reflecting the cubic symmetry of the ferritin shell (Figure 6.19d).

The proposed mechanism for the growth of ferrihydrite is as follows.

1   Iron is delivered to ferritin as Fe^{2+} and then oxidised to Fe^{3+} at a specific site in the shell of the polypeptide cage, near the end of the threefold

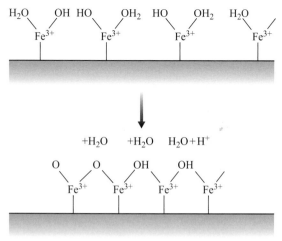

Figure 6.20 Linking of Fe^{3+} ions via condensation at the nucleation site of ferritin.

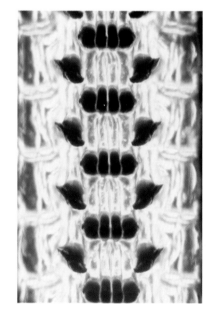

Figure 6.21 Part of a limpet radula showing rows of teeth.

symmetry iron entry channel known as the ferroxidase centre (Figure 6.19a).

2 As more Fe^{2+} enters the cavity and is oxidised, Fe^{3+} ions migrate inwards. Because these Fe^{3+} ions are now close together, they become linked via O or OH bridges, see Figure 6.20.

3 Further Fe^{3+}, OH^- and O^{2-} ions then build on this nucleus to form the iron store.

4 This results in the formation of closely packed polycrystalline structures of ferrihydrite (Figure 6.19c).

6.7.2 Goethite and lepidocrocite

The iron hydroxide oxides goethite and lepidocrite are found in the teeth of limpets and chitons, respectively. Limpet and chiton teeth occur as an array of rows, about 17 teeth per row for chitons and less for limpets arranged on a tongue-like organ known as a radula. A magnified image of part of the radula of a limpet is shown in Figure 6.21.

Teeth are formed at the back of the radula and mature as they move forward. When the animals feed, the radula is pushed forward and the front teeth are used to scrape algae from rocks. The front teeth are gradually eroded and are replaced by new teeth forming at the back. A recent study (Shaw et al., 2008) has measured the rate of production of limpet and chiton teeth at around 0.5 rows per day. Previous studies on gastropods (snails), which also have radula, but whose teeth do not contain iron oxides, gave a rate of production of 1–6 rows per day. Thus it would appear that the iron oxides of the limpet and chiton teeth make these teeth more robust, although there are other factors that affect the production rate such as tooth size and shape and the properties of the food source.

Goethite and lepidocrocite are produced by dissolution of ferrihydrite and slow crystallisation.

■ Why is slow crystallisation important?

☐ Under conditions of rapid crystallisation, the product is kinetically controlled and is likely to be ferrihydrite rather than goethite or lepidocrocite.

Direct conversion from a solid amorphous phase into an ordered phase does not proceed because there is little correlation between the ordered and disordered structures.

The shape of the goethite crystals in limpet teeth is determined by an array of filaments and tubules of chitin, a polysaccharide. This array forces the crystals to grow preferentially along the z-axis and restricts their width to 30–50 nm.

6.7.3   Magnetotactic bacteria

Magnetotactic bacteria use magnets to locate the sediment–water interface in ponds and the sea. The bacteria prepare a chain of nanometre-scale phospholipid vesicles inside which crystals of the magnetic iron oxide, magnetite (Fe_3O_4), form. Iron enters the bacteria as chelated Fe(III), which is then reduced to Fe(II) as it enters the cell. It is then reoxidised to Fe(III) and precipitates as ferrihydrite inside the vesicles. A proportion of Fe(III) ions on the surface of the ferrihydrite are reduced to Fe(II) and these dissolve. Aqueous Fe^{2+} ions bind to the surface of the ferrihydrite via loss of protons from surface OH groups to give Fe(II)–O bonds. Mixed Fe(II)/Fe(III) hydroxo complexes are then lost from the surface to the solution. These lose protons and precipitate as Fe_3O_4. Over time, all the ferrihydrite is converted into magnetite in the form of a single crystal. The process is shown in Figure 6.22.

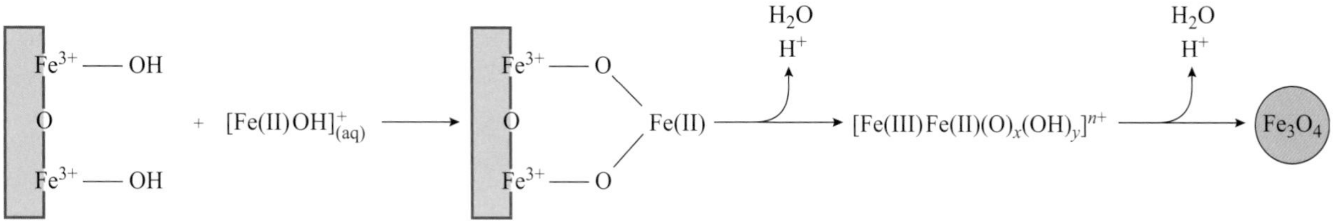

Figure 6.22   Formation of magnetite in magnetotactic bacteria.

You saw earlier (Section 6.2.3) that there was a net alignment of spins in crystals of magnetite. Large crystals, however, are not necessarily magnetic. In large crystals, there are domains separated by grain boundaries (Section 6.3). Within domains, there is a net alignment of spins. However, in a large crystal there are many domains and the spins in different domains can align in different directions. Application of a strong magnetic field or hammering the heated material can force spins in different domains to align and large crystals that act as permanent magnets can be produced this way. If a crystal is small enough, however, it will consist of a single domain so that all the unpaired spins in the crystal are aligned. The crystals produced by magnetotactic bacteria are of a size that falls into the single domain region. They consist of an elongated single domain produced by preferential growth along the body diagonal. Interestingly this direction is that along which it is most favourable for the unpaired spins on iron in magnetite to be aligned. Thus the magnetite in these bacteria is always magnetic.

Don't forget that there are questions on the companion website which you can use to test your understanding of the material covered in this chapter.

7 The role of metal-containing proteins in biological processes

Metals play a wide variety of roles in biological processes. You have met some of them in previous chapters, for example the use of alkali and alkaline-earth metal ions to trigger cell responses and calcium salts for building structures such as bone and shell. In this chapter and Chapter 8 we concentrate on metalloproteins, proteins that have a metal ion as an integral part of their structure. The roles of metals in metalloproteins include:

1 determining or maintaining structures

2 acting as a catalytic site for reactions

3 transferring atoms or groups to catalytic sites

4 transferring electrons for oxidation/reduction reactions

5 storing and transporting molecules.

Here we illustrate some of these, using a number of examples chosen to emphasise the variety of roles that metals can play in metalloproteins.

We start by considering the structural role of metals. You saw that in biomineralisation (Chapter 6) the attachment of metal ions to amino acid residues on an adjacent protein could form a template for crystallisation. In metalloproteins, metal ions can, through coordination, determine the local structure of the protein and even stabilise the protein folding. An important example of a metal in a structural role is zinc in the zinc-finger proteins. This will be our first example.

7.1 Metal ions in a structural role – zinc-finger proteins

The human adult's recommended daily intake of zinc is almost the same as that of iron and zinc occurs in a wide range of proteins. The occurrence of zinc(II) in a protein's structure is now known to be widespread in biochemistry; it is an essential structural element of many proteins.

Zinc-finger proteins have a very precise and rigid higher-order structure, which is important in their function. The main function of zinc-finger proteins is to recognise and bind to DNA. The zinc-finger proteins work by wrapping around the DNA double helix; they have 'finger-like' projections, which fit inside the grooves of the helix (Figure 7.1). The importance of this is that the proteins can recognise specific base sequences of a DNA strand, because the stereochemistry of the finger-like projections places certain amino acid residues in the correct conformation to form hydrogen bonds with the required sequence of nucleic acid bases.

7.1.1 How zinc maintains the protein fingers

Single-crystal X-ray crystallography of a zinc-finger protein reveals that zinc is present in the proteins as a structural element. In fact, zinc is needed to

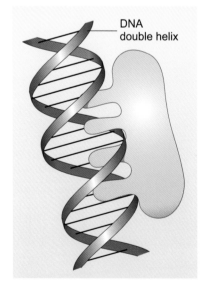

Figure 7.1 Schematic diagram of zinc-finger protein interacting with the double-helix structure of DNA.

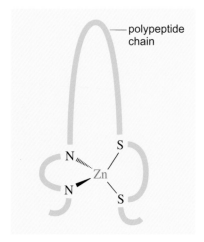

Figure 7.2 Schematic representation of the zinc-finger structure: N represents a histidyl side chain and S represents a cysteinyl side chain.

maintain the finger-like structure of the protein. Figure 7.2 shows how this is achieved. The zinc ion is actually at the 'base' of the 'finger', where it is coordinated by four amino acid side chains: two histidyls and two cysteinyls. The rest of the finger is a loop of polypeptide chain. If the zinc were not present, the finger shape would no longer be pinned at its base; it would then collapse, with subsequent loss of protein activity.

■ Zinc(II) is classed as a 'borderline' acid in hard/soft acid–base theory, and, as such, will form stable complexes with a variety of ligand types. Is this borne out by the structure in Figure 7.2?

☐ The zinc is coordinated to both a soft ligand (cysteinyl) and a hard/borderline ligand (histidyl).

In Figure 7.2, the coordination around zinc is roughly tetrahedral. Is this what would be expected?

In the Periodic Table, zinc is sandwiched between the first-row transition metals and Groups 13–18 of the fourth-Period main-Group elements.

■ Is zinc classed as a transition element?

☐ Not usually. The criterion for a transition element is that the atom has a partially filled d subshell. The only important oxidation state of zinc is zinc(II), which has the electronic configuration $[Ar]3d^{10}$.

So zinc compounds exhibit one major oxidation state, +2. The +2 state is stable, as in this oxidation state the zinc(II) ion has a full ten electrons in its 3d orbitals. The lack of other stable oxidation states is an advantage for the function of zinc-finger proteins as it removes the danger of the metal ion oxidising DNA.

■ What is the crystal-field stabilisation energy (CFSE) for zinc(II) in tetrahedral and octahedral coordination geometries?

☐ Zinc(II) has no CFSE in either the tetrahedral or octahedral case, as its d orbitals are fully occupied. Hence neither geometry is strongly favoured by CFSE.

In practice, the coordination complexes of zinc are found mostly to have four-coordinate tetrahedral geometry, although some five- and six-coordinate complexes are known.

7.1.2 Spectroscopic studies

As we mentioned above, removal of zinc from zinc-finger proteins inactivates them for DNA-binding. That zinc is needed to maintain the finger-like structure was confirmed by an important experiment, in which a zinc-free protein was reactivated by the addition of a cobalt(II) salt (Figure 7.3).

■ Would you expect zinc(II) complexes to be either coloured or magnetic?

☐ With a configuration of $[Ar]3d^{10}$, we would expect the complexes normally to be colourless and diamagnetic. If any zinc(II) compound is

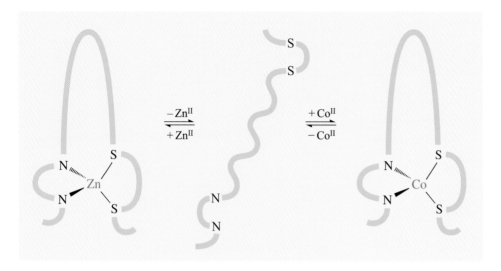

Figure 7.3 Schematic representation showing how zinc(II) is an essential structural feature of zinc-finger proteins, and how the finger structure can be reconstructed with cobalt(II).

coloured (for example, zinc(II) selenide, ZnSe, is orange), the colour comes from charge-transfer or ligand-to-ligand electronic transitions, as no d–d transitions occur in the fully occupied d^{10} configuration.

As most zinc complexes are colourless and diamagnetic, they are referred to as **spectroscopically silent**, and spectroscopic investigation (other than X-ray) of zinc(II) complexes does not give much information.

Cobalt(II) is a good replacement for zinc(II) because the d^7 configuration of cobalt(II) gives relatively stable tetrahedral complexes.

■   Calculate the CFSE of tetrahedral iron d^6, cobalt d^7 and nickel d^8 complexes. Which configuration is the most stable?

☐   The CFSEs are: iron(II), $\frac{3}{5}\Delta_t$; cobalt(II), $\frac{6}{5}\Delta_t$; and nickel(II), $\frac{4}{5}\Delta_t$, suggesting d^7 configuration of Co(II) complexes are the most stable.

Cobalt(II) replaces the zinc(II) ion, forming a tetrahedral complex, as in the zinc(II) protein, and the finger structure is reformed. In fact, before a crystal structure of the protein was available, the tetrahedral coordination geometry of zinc(II) in this protein was confirmed by replacing the zinc(II) in the protein with cobalt(II).

■   How does cobalt(II) differ from zinc(II), and therefore what method might we use to investigate it?

☐   Cobalt(II), unlike zinc(II), has partially filled d orbitals (d^7), which exhibit d–d transitions. These transitions can be observed in the visible spectrum of Co(II) complexes.

When cobalt(II) is substituted into a zinc-finger protein, the resulting visible spectrum shows cobalt d–d transitions with large molar absorption coefficients (greater than 500 $dm^3 \, mol^{-1} \, cm^{-1}$).

■ Why does the high molar absorption coefficient suggest that the coordination geometry of cobalt(II) in the zinc-finger protein is tetrahedral and not octahedral?

□ The two possible d-electron splitting diagrams for tetrahedral and octahedral (high-spin) cobalt(II) are shown in Figure 7.4. In the octahedral complex, any d–d transition is theoretically disallowed by the selection rule, which forbids transitions from a g level to a g level and from a u level to a u level in complexes with a centre of symmetry. Therefore we would only expect to see a very weak absorption band for a d–d transition. On the other hand, d–d transitions in tetrahedral complexes are not forbidden by this rule, and should have a strong absorption band in their UV/visible spectra.

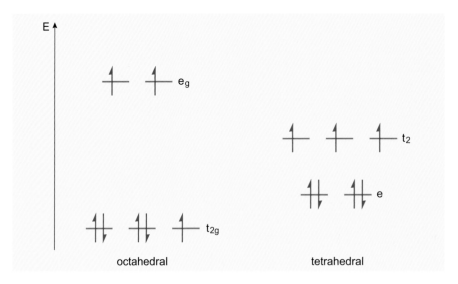

Figure 7.4 Crystal field diagrams for octahedral and tetrahedral high-spin cobalt(II).

From this experiment, it was concluded that the coordination environment in cobalt(II) substituted zinc-finger proteins was probably tetrahedral, and, by analogy, the natural zinc form of the protein would contain tetrahedral zinc (II). This was later confirmed, as you have seen, by single-crystal X-ray diffraction.

7.2 Metals as catalytic centres – liver alcohol dehydrogenase

The second role we consider is that of metal ions as catalytic sites.

■ How do catalysts such as enzymes affect reactions?

□ Catalysts (including enzymes) alter the rate of reaction. The rate constant for a reaction can be written $k = A \exp(-E_a/RT)$. The catalyst can alter the rate of reaction by increasing the A factor and/or decreasing the activation energy, E_a.

Metal ions play a number of different roles at the catalytic site.

Metal ions can act by coordinating reactants so that they are held close together. This will effectively increase the A factor. In some cases, the orientation of the reactants and hence the stereochemistry of the products may be determined by coordination to the metal.

Metal ions can stabilise reaction intermediates, thus decreasing E_a. They can do this by acting as Lewis acids, accepting electrons from reactants and hence stabilising the formation of intermediates produced by the attack of nucleophiles. In this section you will study an example of this – a zinc centre in the enzyme **liver alcohol dehydrogenase**, **LADH**, which plays a role in the metabolism of alcohol.

In other enzymes, stabilisation of the intermediate is accompanied by a change in oxidation state of the metal. An example of this is Cu–Zn superoxide dismutase (SOD) in which O_2^- coordinates to a Cu(II) ion, reducing the copper to Cu(I). You met this enzyme in Chapter 2 and will study it in more detail in Chapter 8.

7.2.1 The function of LADH

The function of LADH is to catalyse the oxidation of primary alcohols to aldehydes (or secondary alcohols to ketones). For example:

$$CH_3CH_2OH = CH_3CHO + H^+ + H^- \tag{7.1}$$
$$\text{ethanol} \qquad \text{ethanal}$$

The products of this reaction include a hydride ion, H^-, which is normally very strongly reducing, and is potentially detrimental to the cell. During the LADH catalytic cycle, the hydride is not generated directly, but is transferred to a cosubstrate, nicotinamide adenine dinucleotide, NAD^+.

The overall reaction catalysed by LADH (where deprotonated ethanol (bound to the enzyme) is used as an example) is:

$$\text{(7.2)}$$

or more simply:

$$CH_3CH_2O^- + NAD^+ = CH_3CHO + NADH \tag{7.3}$$

LADH is important in the metabolism of alcohol in the body and therefore is the centre of much research.

The crystal structure of LADH has been determined, and has been found to contain two types of zinc(II) ion. In one of these, the zinc is coordinated by four cysteinyl side chains (Figure 7.5a), and has no immediate role in the catalytic action of the enzyme. It is another example of zinc playing a structural role. This zinc has an essential role in stabilising the overall higher-order structure of the enzyme. The other zinc ion is coordinated by two cysteinyl side chains, one histidyl side chain and one water molecule; it is this zinc that is essential for catalytic activity (Figure 7.5b).

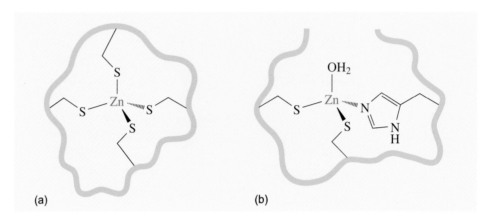

(a)   (b)

Figure 7.5 Close-up of the two zinc sites in LADH: (a) the zinc with a structural role (the cysteinyl residues are shown as 'S'); (b) the zinc that takes part in the catalytic cycle.

7.2.2 LADH as a catalyst

Box 7.1 Catalytic cycles

A catalyst alters the rate of a reaction but although it interacts with the reactants, it is not itself a reactant and is restored to its initial state at the end of the reaction. Reaction schemes for catalysed processes are often shown as catalytic cycles. Starting with the catalyst in its initial or resting state, shown at the top of the cycle, the diagram indicates the addition of substrate and its interaction with the catalyst. The cycle then follows the changes in the catalyst during intermediate reactions in the process ending with the catalyst returning to its initial state. In this representation, states of the catalyst are linked by arrows, with reactants appearing on arrows leading into the cycle, and products appearing on arrows leading out of the cycle.

The catalytic cycle (see Box 7.1) of LADH is shown in Figure 7.6. It shows some important features. In step A the NAD^+ binds adjacent to the zinc site. In step B the alcohol substrate displaces the water from the zinc. In step C, the Lewis acidity of the zinc (Box 7.2) facilitates the deprotonation of the bound alcohol to give a bound alkoxide. The negative charge of the alkoxide is lost by the rapid transfer of H⁻ to NAD⁺ (step D). In the final step, step E, the ethanal is displaced by water, NADH leaves the active site, and the cycle begins again.

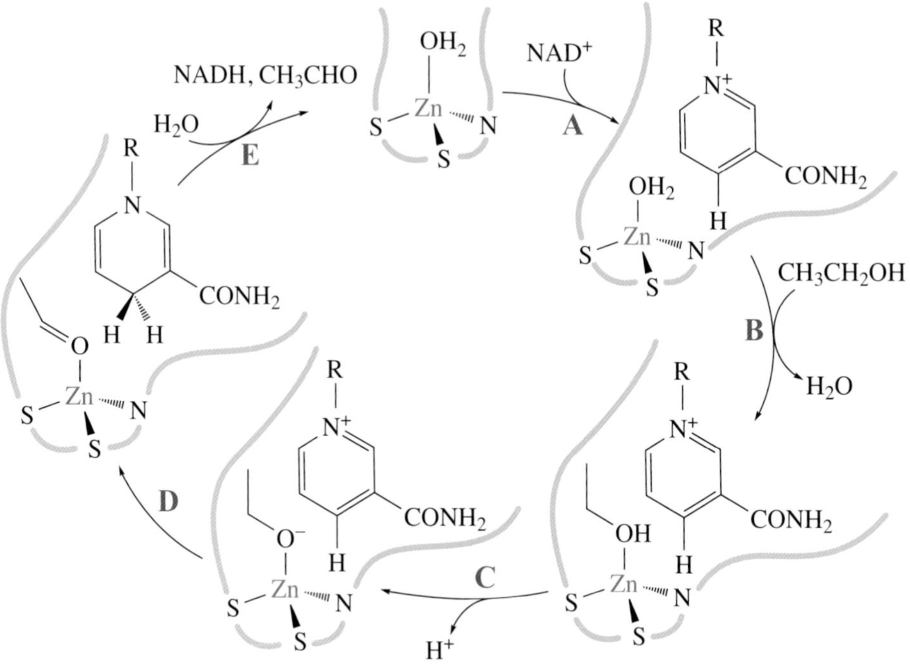

Figure 7.6 Catalytic cycle of LADH, using ethanol as an example; the cysteinyl residues are represented as 'S' and the histidyl residue as 'N'.

Box 7.2 Zinc as a Lewis acid

Lewis acids are electron pair acceptors. Zinc(II) is a reasonably strong Lewis acid. In organic synthesis, zinc salts are often used as catalysts, where the zinc acts as a Lewis acid. For example, in a Friedel–Crafts alkylation of a benzene ring, $ZnCl_2$ is sometimes used as a catalyst:

$$C_6H_5R + R'Cl \xrightarrow{ZnCl_2} C_6H_4RR' + HCl \qquad (7.4)$$

The mechanism of this reaction is shown as steps 7.5–7.7; the zinc(II) polarises the attacking substrate, and thereby enhances its electrophilicity:

$$R'-Cl + ZnCl_2 \rightleftharpoons \overset{\delta+}{R'}-Cl \rightarrow \overset{\delta-}{ZnCl_2} \qquad (7.5)$$

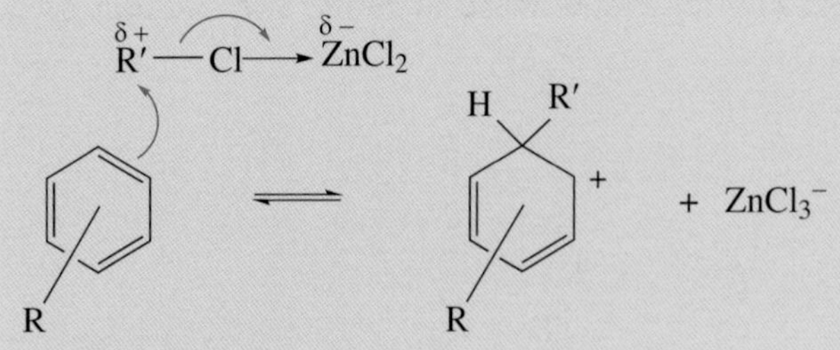

$$\text{Zn}^{2+}(\text{aq}) + \text{H}_2\text{O}(\text{l}) = [\text{Zn(OH)}]^+(\text{aq}) + \text{H}^+(\text{aq}) \tag{7.6}$$

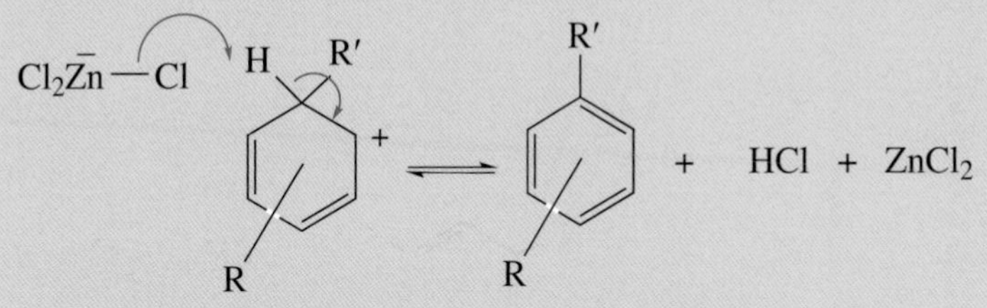

Equations (7.6) and (7.7) shown in the diagram box above.

Zn^{2+} will also polarise coordinated water molecules, and zinc(II) salts dissolved in water will undergo hydrolysis:

$$\text{Zn}^{2+}(\text{aq}) + \text{H}_2\text{O}(\text{l}) = [\text{Zn(OH)}]^+(\text{aq}) + \text{H}^+(\text{aq}) \tag{7.8}$$

$$[\text{Zn(OH)}]^+(\text{aq}) + \text{H}_2\text{O}(\text{l}) = \text{Zn(OH)}_2(\text{s}) + \text{H}^+(\text{aq}) \tag{7.9}$$

■ How does the Lewis acidity of zinc(II) in LADH help the deprotonation of the alcohol?

☐ The zinc(II) polarises the alcohol so that the H of the OH group has a larger partial positive charge, $\delta+$, and hence is more easily lost as H^+ (step C).

7.3 Group transfer to a catalytic site – vitamin B_{12}

As well as being present at the active site of a metalloprotein, metal ions can also be used to transfer reactants to the active site. One such example is illustrated by **vitamin B_{12}**.

The discovery that patients with pernicious anaemia could be treated by being given large amounts of raw liver was a breakthrough in the management of this disease, but not a solution that was attractive to patients. The active ingredient in liver was found to be vitamin B_{12}, which is present in meat, fish, eggs and milk – but not in fruit or vegetables. Most people ingest sufficient amounts of this vitamin with their food. However, it is possible to develop an autoimmune disease in which the body does not absorb vitamin B_{12} properly and this leads to pernicious anaemia.

The crystal structure of vitamin B_{12} was solved by 1956, work for which Dorothy Crowfoot Hodgkin won the Nobel Prize in 1964. The structure is shown in Figure 7.7.

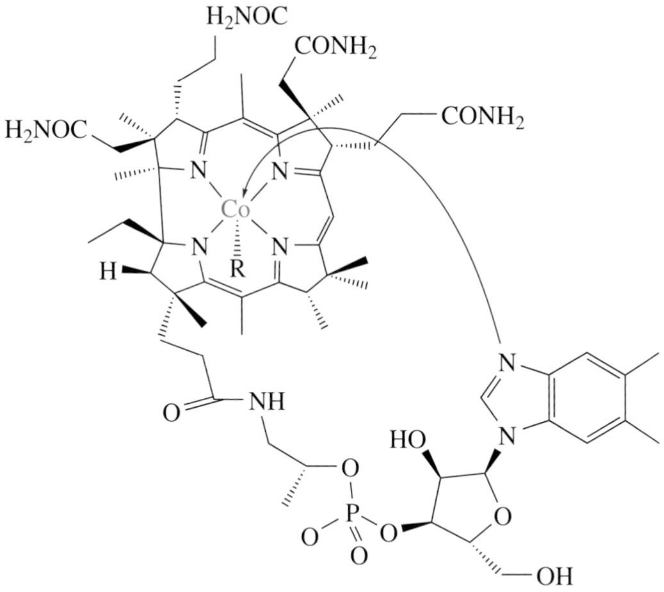

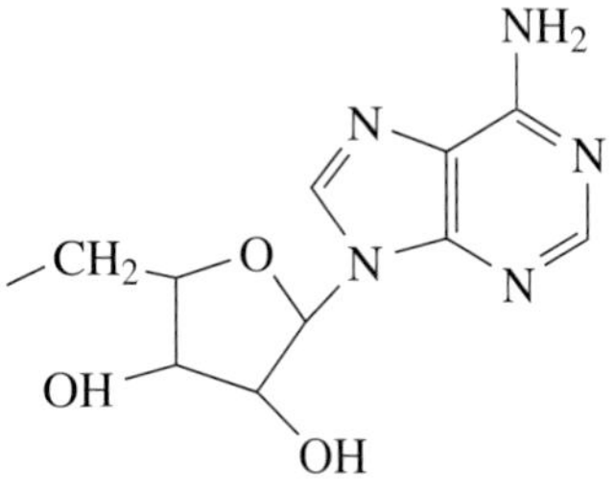

Figure 7.7 Structure of vitamin B_{12}. In the original determination, the extraction process produced a CN group attached to cobalt. However, in humans, the ligand R is either a methyl group or a 5′-deoxyadenosyl group, **7.1**.

At its centre is a cobalt ion surrounded by a four-ring structure reminiscent of the porphyrin ring in haem, known as a **corrin**. However, although there are four rings coordinating to cobalt via N, there are important differences between the corrin and the porphyrin structures. The porphyrin group has a delocalised structure of alternate single and double bonds, whereas in the corrin group this delocalisation is interrupted because two of the five-membered rings are linked directly rather than through an intervening carbon. One consequence of this is that the corrin ring has nine carbons that are **chiral** centres, whereas the carbons of the porphyrin group are non-chiral.

Coordinating cobalt, perpendicular to the corrin plane, are a nitrogen from a nucleotide derived from dimethylbenzimidazole and a carbon from either a methyl group or a more complex 5′-deoxyadenosyl group. The name vitamin B_{12} is applied to the cyanide derivative, **cyanocobalamin**. The 5′-deoxyadenosyl compound is known as **coenzyme B_{12}** and the methyl compound as **methylcobalamin**. It is the methyl or 5′-deoxyadenosyl group that is transferred to the active site of the enzyme and reacts with the substrate. To be useful as a group transfer agent, coenzyme B_{12} or methylcobalamin must lose the transferred group easily. We shall now see how these coenzymes fulfil their role.

The presence of a metal–carbon bond in biological systems is rare although compounds with such bonds form a whole area of chemistry – organometallic chemistry. Transition metals form a wide variety of organometallic compounds but most of these involve unsaturated organic ligands. The reason for this is that transition metal–alkyl bonds are generally weak. Thus coenzyme B_{12} and methylcobalamin are unusual for organotransition metal compounds as they do contain metal-alkyl bonds. The Co-5′-deoxyadenosyl bond enthalpy is 130 kJ mol^{-1}, compared to an average C–C bond enthalpy of 347 kJ mol^{-1}.

7.3.1   Mechanism of coenzyme B$_{12}$-catalysed reactions

This weakness in the metal–carbon bond is exploited in the use of coenzyme B$_{12}$. Many of the processes that coenzyme B$_{12}$ helps catalyse involve radical rearrangement reactions. The starting point for these is generally considered to be the formation of the 5′-deoxyadenosyl radical by breaking of the Co–C bond. A series of reactions then follows. The 5′-deoxyadenosyl radical abstracts hydrogen from a substrate molecule to form a substrate radical. The substrate radical rearranges to form a product radical. The product radical then abstracts a hydrogen from 5′-deoxyadenosine, regenerating the 5′-deoxyadenosyl radical and forming the product molecule. The 5′-deoxyadenosyl radical then reattaches to the cobalt, ready to start the cycle again. This is shown in schematic form in Figure 7.8. It should be noted, however, that experiments have failed to detect the 5′-deoxyadenosyl radical.

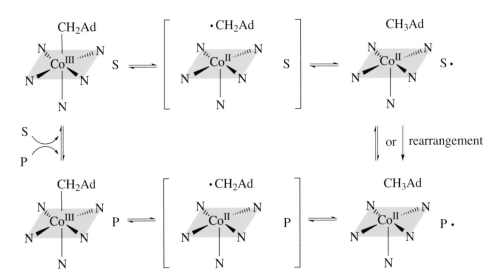

Figure 7.8   General scheme for the mechanism of rearrangement of a substrate, S, to form product, P, catalysed by coenzyme B$_{12}$. CH$_3$Ad represents 5′-deoxyadenosine. It is not clear whether the rearrangement is one way or an equilibrium. Note that the N below the plane represents the nucleotide in Figure 7.7.

You should note that the effect of the weak Co–C bond is a kinetic one. The bond enthalpy can be equated to the activation energy for the production of the 5′-deoxyadenosyl radical. Experiments in which the decrease in intensity of a peak at 525 nm in the electronic spectrum of the coenzyme B$_{12}$ was monitored (Licht et al., 1999) yielded values for the rate constant, k, at various temperatures for the process

$$\text{coenzyme B}_{12} \rightarrow 5'\text{-deoxyadenosyl} + \text{reduced coenzyme B}_{12} \qquad (7.10)$$

As $k = A \exp(-E_a/RT)$, the activation energy, E_a, and the A factor can be obtained from the variation of k with temperature. Coenzyme B_{12} does not however act alone in living systems; it works in conjunction with an enzyme. In the presence of the enzyme, the rate of reaction is of the order of 10^{10}–10^{11} faster than that measured for Equation 7.10.

■ How might the enzyme cause this rate increase?

☐ By increasing the A factor or decreasing the activation energy, E_a.

It is thought that the enzyme lowers the activation energy

■ The Co–5′-deoxyadenosyl bond energy is 130 kJ mol^{-1}. Assuming the reaction takes place at 37 °C (310 K) and that the A factor does not change, what value must E_a be reduced to in order to increase the rate constant by a factor of 10^{10}? [*Hint*: $\ln k = \ln A - (E_a/RT)$]

☐ Let the enzyme-free rate constant be k and that for the reaction in the presence of the enzyme be k'. Then $\ln k' - \ln k = -E'_a/RT + E_a/RT$.

By the division rule for logarithms, $\ln k' - \ln k = \ln(k'/k)$.

$(k'/k) = 10^{10}$, thus $\ln(k'/k) = \ln(10^{10}) = (E_a - E'_a)/RT$

$(E_a - E'_a) = RT \ln(10^{10})$

$E'_a = E_a - RT \ln(10^{10})$

$E'_a = (130\,000 - 8.314 \times 310 \times 23.03)$ J mol^{-1} = 70.6 kJ mol^{-1}

Thus the Co–5′-deoxyadenosyl bond enthalpy needs to be reduced to about 70 kJ mol^{-1}.

There are several ways in which the enzyme could do this – it could stabilise the Co(II) form of the coenzyme that is produced as the bond breaks by reducing the electron-donating ability of the axial ligand or by increasing the Co–axial ligand bond distance; it could also lead to deformation of the corrin ring in such a way that there was steric interference between ring substituents and the 5′-deoxyadenosyl ligand.

7.3.2 Vitamin B_{12} in the body – formation of succinyl–CoA

There are two important reactions in humans that are catalysed by vitamin B_{12} derivatives. One is the rearrangement of methylmalonyl–CoA, **7.2**, to succinyl–CoA, **7.4**. (These are derivatives of methylmalonic acid, **7.3**, and succinic acid, **7.5**.) This rearrangement is important because the biosynthesis of all tetrapyrroles, such as haem, starts from 5-aminolaevulinic acid, **7.6**, which in animals is formed from glycine and succinyl–CoA. Thus faults in cobalt metabolism can lead to anaemia even if iron intake is adequate.

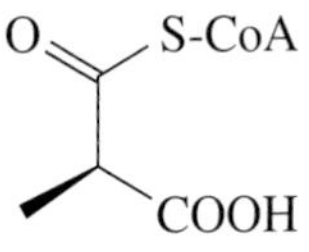

7.2 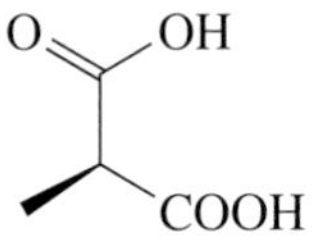   7.3 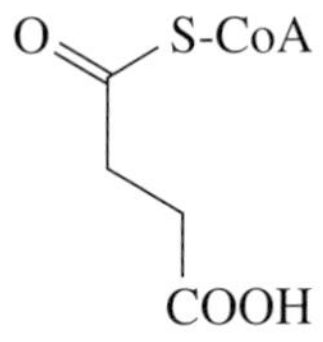   7.4

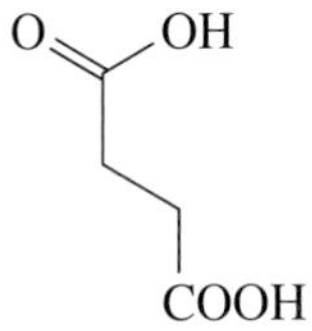

7.5 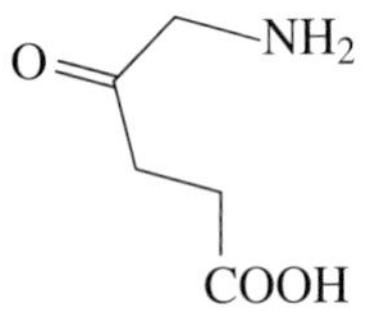   7.6

Possible mechanisms of this rearrangement are shown in Figure 7.9. It has been proposed that the basic group (BH) hydrogen-bonding to the carbonyl oxygen is a histidine residue.

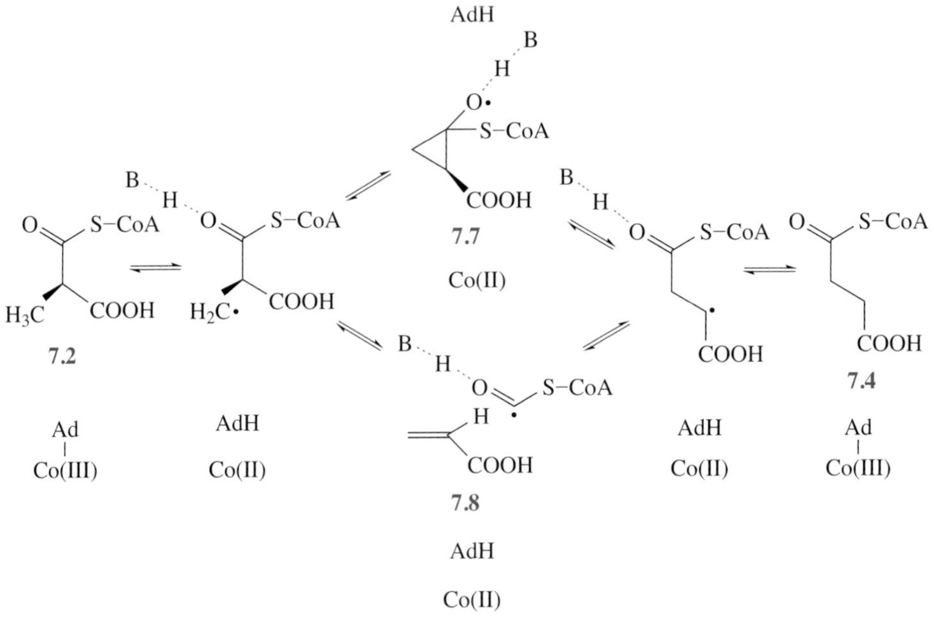

Figure 7.9   Mechanisms of conversion of methylmalonyl–CoA (**7.2**) into succinyl–CoA (**7.4**). (Ad = 5′-deoxyadenosyl.) (Adapted from Banerjee (2003).)

5′-deoxyadenosyl is released from the coenzyme B_{12} and gains a proton from the methyl group of methylmalonyl–CoA. The deprotonated methylmalonyl–CoA can rearrange to deprotonated succinyl–CoA via an intermediate. The nature of the intermediate is uncertain and Figure 7.9 shows two possibilities.

Finally, 5'-deoxyadenosyl donates a proton to the succinyl–CoA radical and re-attaches to the cobalamin.

The resting state of the coenzyme contains Co(III).

■ The resting state is diamagnetic. What does this suggest about the crystal field in this state?

☐ Co(III) is a d^6 ion. The environment in the coenzyme can be regarded as approximately octahedral. The observation that it is diamagnetic suggests that it is a low-spin (strong-field) t_{2g}^6 complex.

When the Co–C bond is broken, the cobalt is reduced to Co(II). The remaining ligands around Co form a square pyramid and the d orbitals will be split as in Figure 7.10. In the reduced coenzyme B_{12}, Co(II) has only one unpaired electron. This is a low-spin complex with an empty $d_{x^2-y^2}$ orbital and one electron in d_{z^2}.

■ Suggest why a low-spin Co(II) complex might be square pyramidal rather than octahedral.

☐ Low-spin d^7 octahedral complexes have one electron in an e_g orbital and are therefore in a degenerate state. According to the Jahn–Teller theorem, such complexes will distort to remove the degeneracy. Square-pyramidal complexes can be thought of as very distorted octahedral complexes.

7.3.3 Vitamin B_{12} in the body – production of methionine

The second important reaction in humans catalysed by vitamin B_{12} derivatives is the conversion of homocysteine, **7.9**, into the essential amino acid, methionine. This involves methylcobalamin. Methionine is used to produce S-adenosylmethionine, SAM, a molecule used as a methylating agent in the body, for example to methylate DNA and regulate gene expression.

SH

H₂N

NH₂

O

OH

7.9

H₂N, N, H, N, NH, N, O, CH₃, H, NHR

7.10

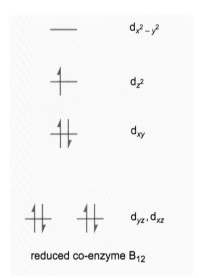

reduced co-enzyme B_{12}

Figure 7.10 d electron energy levels for a square-pyramidal arrangement of ligands and electron occupancy as in Co(II) in reduced coenzyme B_{12}.

The conversion of homocysteine into methionine is complex. The catalytic system involves the enzyme methionine synthase and an additional cofactor, N^5-methyltetrahydrofolate, **7.10**, as well as methylcobalamin. Deficiencies of both vitamin B_{12} and folic acid can thus interfere with this process. Here we shall focus on the methylcobalamin. As with coenzyme B_{12}, there is cleavage of the Co–C bond. In contrast to the homolytic cleavage of the Co–5'-adenosyl bond, the cleavage of the Co–Me bond is heterolytic. The methyl group attaches to homocysteine replacing the proton on sulfur and the cobalt is reduced to Co(I).

■ What is the d electron configuration of Co(I)?

□ d^8.

■ What geometry would you expect for a transition metal d^8 ion?

□ d^8 ions such as Ni^{2+} and Pt^{2+} often form complexes with square-planar geometry. In this geometry, the $d_{x^2-y^2}$ orbital is much higher in energy than the other d orbitals. A d^8 ion has enough electrons to fill the lower orbitals and leave this one empty (Figure 7.11).

Structural studies have indicated that the Co(I) in reduced cobalamin is indeed in a four-coordinate environment. It has been proposed that the ligand opposite methyl (see Figure 7.7) is protonated on reduction to Co(I) and therefore the N atom is no longer coordinated to Co.

Co(I) complexes are quite unusual.

■ What are the most common oxidation states of Co in its complexes?

□ +2 and +3.

In simple halide and oxide compounds, oxidation states of +4 and +5 have also been observed but not +1.

The $(Co^{2+}|Co^+)$ couple is negative so that reduction to Co^+ is thermodynamically unfavourable. This instability of Co(I) with respect to Co(II) leads to an undesirable side reaction. The Co(I) cobalamin can react with O_2 to form Co(II) cobalamin and superoxide radicals. Co(II) cobalamin is inactive as a cofactor and as we saw in Section 3.2 superoxide radicals are hazardous. Normally the Co(I) complex would react rapidly with N^5-methyltetrahydrofolate to regenerate methylcobalamin. However about once in every 2000 cycles, the side reaction occurs and a self-repair system has evolved to deal with this. Part of the enzyme methionine synthase has adapted to catalyse the methylation of Co(II) cobalamin with *S*-adenosylmethionine, the universal methylating agent.

7.4 Electron transport

There are many biological processes that involve electron transfer and in this section we will discuss some of the proteins specialised for this purpose. You will meet these proteins again as part of the processes discussed in Chapter 8. The transfer of electrons from one protein to another is essentially an oxidation/reduction reaction. First let us see what the requirements for a good electron transfer system are.

1 The biological processes usually involve one-electron transfer – that is, we need a system with two oxidation states separated by one.

2 The two states should have geometries that are similar to minimise the energy needed to reorganise atoms.

3 There should be efficient pathways between the electron donor/acceptor site and the site that is oxidised/reduced.

square-planar complex

Figure 7.11 Crystal field energy diagram for a d^8 ion in a square-planar complex.

4   Electron transfer is faster if the free energy change for the reaction in which an electron is transferred from the donor to the acceptor is small and negative. The overall process can be written in two steps (where ETP is electron transfer protein)

reduced D + oxidised ETP → oxidised D + reduced ETP

reduced ETP + oxidised A → oxidised ETP + reduced A

where D is a protein acting as an electron donor and A is a protein acting as an electron acceptor.

The electrode potential for the couple (oxidised ETP|reduced ETP) thus needs to be between that for the electron donor and that for the electron acceptor.

Most electron transfer proteins contain iron or copper.

■   List some reasons why proteins containing Fe or Cu might make good electron transfer proteins.

☐   (i) Both Fe and Cu have two stable oxidation states separated by one, (Fe(III)/Fe(II) and Cu(II)/Cu(I)).

(ii) It is possible, especially with Fe, to prepare complexes of the two oxidation states with similar geometries.

(iii) The electrode potentials can be varied by changing the ligands round the metal and so it is possible that a careful choice of ligating groups will produce an active site with the required electrode potential.

We shall consider three very different electron transfer systems and see how the properties of Fe and Cu are used. The systems are the cytochromes c, iron–sulfur proteins (ferredoxins) and blue copper proteins.

7.4.1   Cytochromes c

Cytochromes c are found throughout the body and play an essential role providing electrons to many different processes. The active site for all cytochromes c (apart from cytochrome c′) is an Fe ion surrounded by six ligand atoms. Four of the coordination positions are occupied by a porphyrin group. The fifth position is occupied by the N of a histidine residue and the sixth position by the N of another histidine residue or an S of a methionine residue, Figure 7.12.

So how do cytochromes c meet the requirements we listed above?

Well, first they contain an Fe ion which cycles between oxidation states +2 and +3 during oxidation/reduction and so they fit the first requirement.

Second, are there geometry changes when the oxidation state changes? There have been X-ray studies on the oxidised and reduced forms of both yeast cytochrome c and mitochondrial horse cytochrome c. The conclusions were that the bulk of the protein is unchanged by the change of oxidation state, but

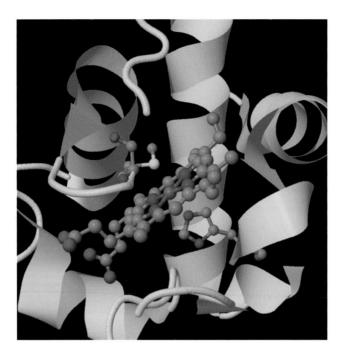

Figure 7.12 Coordination of Fe in a typical cytochrome c. (Based on pdb file 3dr0 (Bialek et al., 2009).)

that there were, however, small differences in the region surrounding the iron. These cytochromes c have a methionine as the sixth ligand coordinating iron and there are changes in the orientation of this ligand. In addition, the haem group itself is not truly planar in either form and the deviation from planarity increases slightly in the oxidised form. Finally a water molecule bound near the active site was observed to move 160 pm nearer the Fe ion in the oxidised form. It was suggested that this water could be oriented so that the negative end of its dipole pointed towards Fe and partially compensated for the positive charge on the haem group in the Fe(III) form. So there are changes in the geometry but they are minor ones that would not lead to a large change in energy.

With respect to the third requirement, insight into the electron transfer path has come from structural studies of yeast cytochrome c/cytochrome c peroxidase complexes. Cytochrome c peroxidase is an enzyme that oxidises cytochrome c in yeast. The X-ray structure of the complex of cytochrome c and cytochrome c peroxidase has been determined. Part of this is shown in Figure 7.13. You will note that cytochrome c peroxidase also has a haem group. We can use the X-ray structure to illustrate how an electron pathway can be set up to transfer electrons from cytochrome c to cytochrome c peroxidase. The cytochrome c haem group is situated close to the outside of the protein. In the complex, the two proteins interact through electrostatic interactions between positively charged surface residues on cytochrome c and negatively charged residues on cytochrome c peroxidase and through non-bonded interactions. The two Fe ions are relatively close (2650 pm apart) with haem groups and residues held together by non-bonded interactions linking the two.

The proposed pathway is illustrated in the section of the X-ray structure shown in Figure 7.13. This figure shows the non-bonded or van der Waals radii of the atoms. The haem group and the protein chain are in van der Waals contact and the electron can transfer via non-bonding interactions from one haem to the other via the intermediate residues. However, the nature of the pathway is still controversial.

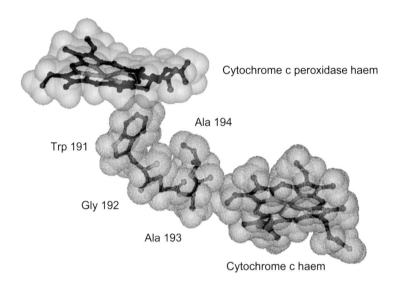

Cytochrome c peroxidase haem

Ala 194

Trp 191

Gly 192

Ala 193

Cytochrome c haem

Figure 7.13 Proposed pathway for electron transfer from cytochrome c to cytochrome c peroxidase. (Based on data from pdb file 2pcb (Pelletier, 1992).)

Finally, we need to consider the electrode potentials for the system. To do this we need to think about what we mean by electrode potential in biochemical systems.

7.4.2 Electrode potentials in biochemical systems

Biological molecules are not usually found in solution in the standard conditions we use to quote $E^{\ominus}$. However, measurements are available for electrode potentials in air-saturated water at pH 7. These electrode potentials will be designated E_0'. The E_0' value for the $(H^+(aq)|H_2(g))$ couple, for example, -0.42 V at pH 7, is very different from the standard redox potential ($E^{\ominus} = 0.00$ V).

Let us see how the potentials vary in an important process that involves cytochrome c – respiration. We shall return to the process of respiration in Chapter 8, but for now we shall just concentrate on the electrode potentials involved. The complete mitochondrial electron transfer chain in respiration is complex and contains several steps (Figure 7.14). In the mitochondrial respiratory chain cytochrome c is oxidised by an enzyme called cytochrome c oxidase (CCO) (we will meet this enzyme again in Chapter 8).

Cytochrome bc_1 catalyses the oxidation of a small lipid-soluble molecule, ubiquinol, **7.11**, to ubiquinone, **7.12**, and the reduction of cytochrome c. CCO reduces O_2 and reoxidises cytochrome c.

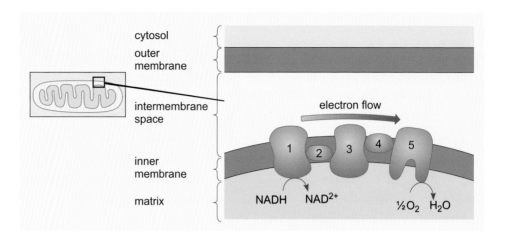

Figure 7.14 The electron transport chain in mitochondria consists of five electron carriers, numbered 1–5 in order of electrode potential. 3 represents an enzyme known as cytochrome bc_1 (also known as Q-cytochrome c oxidoreductase), 4 is cytochrome c and 5 represents cytochrome c oxidase (CCO).

$$ \text{7.11} = \text{7.12} + 2H^+ \qquad (7.11) $$

Thus the electrode potential of cytochrome c should be between that of the (ubiquinol|ubiquinone) couple and that of the (O_2|H_2O) couple.

$$ E_0'(\text{ubiquinone|ubiquinol}) = 0.04 \text{ V} \qquad E_0' (O_2|H_2O) = 0.82 \text{ V} $$

The electrode potential, E_0', of cytochrome c is 0.26 V in solution and 0.23 V when bound to CCO and so fulfils this requirement.

■ $E^\ominus$ for the (Fe^{3+}|Fe^{2+}) couple is 0.77 V. Why is E_0' for cytochrome c so much lower?

☐ There are two points to consider here.

(i) The iron in cytochrome c can be considered to be in a complex and complexing can stabilise one oxidation state relative to the other. For example, $E^\ominus$ for the (Fe^{3+}|Fe^{2+}) couple when Fe is complexed by $CN^\ominus$ is reduced to 0.36 V.

(ii) The electrode potentials quoted above are E_0' values not $E^\ominus$ values and we noted above that the electrode potential, E_0', is considerably lower than $E^\ominus$ for the (H^+(aq)|H_2(g)) couple.

Overall, we can see that the active site in cytochrome c meets all four of the requirements for a good electron transfer protein.

7.4.3 Ferredoxins

There are a large number of proteins with a variety of functions that contain clusters of iron and sulfur atoms. We met one previously in Chapter 5 in the iron regulating proteins and in Chapter 8, you will meet an enzyme containing an iron–sulfur cluster that fixes atmospheric nitrogen, nitrogenase. Here we are concerned with electron transfer proteins, in particular **ferredoxins**. Iron–sulfur proteins represent some of the oldest electron transfer systems in evolutionary terms and are found in species from bacteria to humans. Ferredoxins play an essential part in electron transport chains for respiration, photosynthesis and nitrogen fixation (and may also store iron).

Ferredoxins contain iron–sulfur clusters (centres), containing varying numbers of Fe and S atoms, with a net charge that changes according to the oxidation state of the metal, Fe(II) or Fe(III). Two kinds of sulfur are present in iron–sulfur centres:

1 Inorganic (acid-labile) sulfur. This can be hydrolysed by dilute acid to give H_2S, as in the hydrolysis of iron(II) sulfide:

$$FeS + 2H^+ = Fe^{2+} + H_2S \qquad (7.12)$$

2 Sulfur linking the cluster to the protein, in fact the thiolate (–SH) side chain of cysteine residues. This link is not acid-labile.

Only inorganic sulfur is counted in the description of the cluster. The iron in these proteins is always tetrahedrally coordinated, though with distortions from the ideal geometry. The inorganic sulfur bridges two or three irons.

2Fe–2S, 3Fe–4S and 4Fe–4S clusters all occur in ferredoxins. The geometries of these are shown in Figure 7.15.

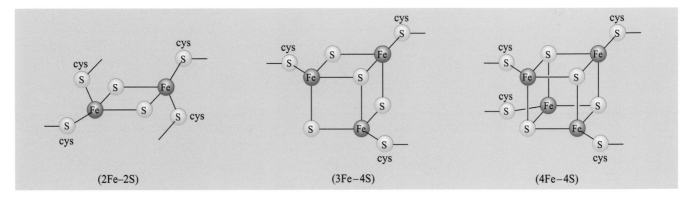

(2Fe–2S)   (3Fe–4S)   (4Fe–4S)

Figure 7.15   Iron–sulfur clusters. Cys is the amino acid cysteine.

Forming a cluster fixes the geometry so that there is little change of structure on oxidation or reduction. There is some spectroscopic evidence that the iron and bridging sulfur form a delocalised system in which the electrons pair up in molecular orbitals. For electron-counting purposes, however, we may say that a particular cluster contains Fe^{2+} (d^6) and Fe^{3+} (d^5) ions.

In living systems, each cluster normally adopts only two oxidation states, differing by one electron; the transport of more than one electron at a time requires more than one cluster in the protein. The clusters are damaged by oxygen, which must be rigorously excluded.

2Fe–2S ferredoxins have two cysteines coordinated to each Fe. The Fe_2S_2 diamond is close to the protein surface and the Fe more exposed to solvent is the one reduced. E_0' values of ferredoxins are large and negative (−0.428 to −0.329 V) making them suitable for electron transfer in photosynthesis. The effect of coordination on E_0' values is illustrated by a group of related proteins known as Rieske proteins. These differ from ferredoxins in having one Fe coordinated by two histidine groups. The change in coordination leads to the E_0' values being less negative (−0.090 to +0.280 V).

4Fe–4S clusters are mainly found in bacterial ferredoxins. This cubane structure is the most stable of the iron–sulfur centres. In these clusters, the Fe–Fe distance of 270 pm is close enough for metal–metal bonding to occur.

Figure 7.16 shows schematically how the clusters are held in ferredoxin in the bacterium, *Clostridium pasteurianum* as determined by X-ray crystallography. The ferredoxin contains two [4Fe–4S] clusters, and Figure 7.16 shows how cysteines attached to two neighbouring irons are commonly separated by two other amino acids (indicated by the numbering of some of the amino acids). Thus, cysteines numbered 35, 38 and 41 are attached to one cluster, and numbers 8, 11 and 14 to the other.

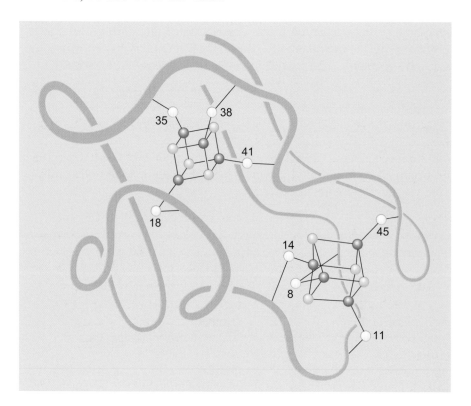

Figure 7.16 Schematic drawing of the polypeptide chains and iron–sulfur centres in ferredoxin from *C. pasteurianum*.

The E_0' values of [4Fe–4S] ferredoxins are, like those of [2Fe–2S] ferredoxins, large and negative.

■ Describe how Fe–S clusters meet the requirements for a good electron transport system given in Section 7.4.

☐ The requirements are met as follows:

1 Fe in the clusters can cycle between oxidation states (II) and (III).
2 The clusters are rigid structures and so there is no change in structure.
3 The pathway has not been discussed.
4 It is stated that the E_0' values are suitable for photosynthesis. (They are also suitable for other processes.)

7.4.4 Blue copper proteins

Blue copper proteins are relatively small soluble proteins found in bacteria and plants. Amongst other functions, they are involved in photosynthesis and respiration. These proteins were named blue because they have an intense peak ($\varepsilon > 3000$ dm^3 mol^{-1} cm^{-1}) at around 600 nm in their visible spectrum which gives them an intense blue colour.

The position of the absorption band suggested that the copper was coordinated to sulfur and this was supported by an EXAFS study of the blue copper protein, **azurin**, which indicated one sulfur atom at 210 pm from Cu. EXAFS also identified two or three nitrogen atoms at 200 pm from Cu. X-ray structures have confirmed that there are two N atoms and one S coordinating Cu at short distances.

The active site of another blue copper protein, **plastocyanin**, is shown in Figure 7.17. Plastocyanin is used as an electron transport protein in photosynthesis.

The copper ion is surrounded by three roughly planar ligands, two histidine residues coordinated through N, a cysteine residue coordinated through S, and a methionine coordinated through S. Other blue copper proteins have a fifth position occupied by a carbonyl group or have the methionone replaced by glutamine which coordinates through O.

The arrangement of ligands around Cu is unusual. We noted that a good electron transfer protein should show very little change in the coordination around the metal atom when going from the reduced to the oxidised state. If we looked at the coordination of Cu in Cu(I) and Cu(II) complexes, we might decide that Cu was not a good candidate for an electron transfer centre because the two oxidation states favour very different coordination. Cu(I), a d^{10} ion like Zn(II), invariably favours a tetrahedral geometry. Cu(II), d^9, complexes adopt distorted octahedral, five-coordinate or square-planar geometries.

■ Why do Cu(II) complexes adopt a distorted octahedral or square-planar geometry rather than a regular octahedral geometry?

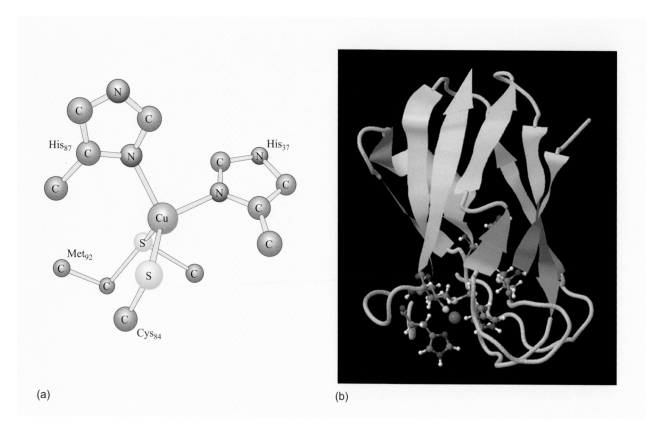

(a)

(b)

Figure 7.17 (a) Active site of plastocyanin showing the environment of the copper. (b) Active site of plastocyanin. (Based on pdb file 1jxd (Bertini et al., 2001).)

□ d^9 ions in a regular octahedral environment would be degenerate as there is a choice of which e_g orbital is doubly occupied. The distorted geometry lowers the energy by removing this degeneracy (Jahn–Teller effect).

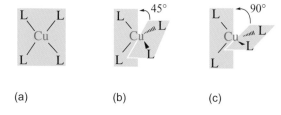

(a) (b) (c)

Figure 7.18 (a) Square-planar copper ion coordination geometry, (b) intermediate copper ion coordination geometry showing two imaginary planes inclined at 45°, and (c) tetrahedral copper ion geometry showing the imaginary planes inclined at 90°.

Tetrahedral geometry would stabilise copper(I) over copper(II). In fact, the coordination geometry in the protein is somewhere between square planar and tetrahedral. This is illustrated in Figure 7.18. Figure 7.18a shows square-planar geometry around a copper ion. A square-planar geometry can be converted into a near tetrahedral geometry by twisting the imaginary plane formed by the copper ion and the coordinating atoms of two adjacent ligands, so that it is at 90° to the plane formed by the copper ion and the other two adjacent ligand atoms (Figure 7.18c). The copper coordination geometry found in these blue copper proteins can be described as a partial rotation of the two imaginary planes, such that the planes are inclined at an angle of about 45° to each other – in other words, half-way between tetrahedral and square-planar geometries (Figure 7.18b). Such a coordination geometry, which is clearly held in place by the rigid higher-order structure of the protein, does not stabilise either of the copper ion's two possible oxidation states with respect to the other. This phenomenon of a metal coordinated by a fixed spatial arrangement of amino acid side chains is called an **entatic state**. Along with the hardness/softness mixture of the amino acid side chains that coordinate to the metal ion, the entatic state is an example of how protein structure controls the electrode potential of the metal ion within the protein.

The geometry around the Cu ion is thus determined by the folding of the peptide chain and remains almost unchanged upon oxidation. By adopting an intermediate entatic state, the protein thus ensures that requirement 2 in Section 7.4 is met.

The copper centre is close to the surface of the protein but not accessible to the surrounding solvent. This will reduce the length of the electron transfer path, whilst preventing solvent attack on the centre.

E_0' values of blue copper proteins range from 0.140 V to 0.700 V. These values are more positive than those of the other electron transfer proteins we have discussed. Thus they are suitable for processes needing greater oxidising power. In photosynthesis, plastocyanin ($E_0' = +0.375$ V) is used for transferring electrons from cytochrome b6f (a complex system with several redox centres whose E_0' values are -0.150 V, -0.050 V, $+0.340$ V and ≈ 0.300 V) to P700 in photosystem I ($E_0' = +0.450$ V). We will meet photosynthesis again in Chapter 8.

In the next chapter, we will look at metalloproteins involved in processes key to life. Respiration provides an example of a metalloprotein used to transport molecules, the final in the list of key roles that we gave at the beginning of this chapter. We will then consider two processes, photosynthesis and nitrogen fixation that, although not occurring in humans, provide us with the carbon and nitrogen that we need. You will find that the electron transfer proteins discussed here are important links in the chains of reactions occurring in these processes.

Don't forget that there are questions on the companion website which you can use to test your understanding of the material covered in this chapter.

8   Some key biological processes

This chapter considers the role of metals in a series of processes essential to life. We start with the uptake, transport and use of oxygen in the human body. How is the oxygen from the air we breathe in transported from our lungs to our muscles and other parts of the body? You will study the oxygen carriers haemoglobin and myoglobin in detail and will look briefly at other oxygen transport proteins.

We will then examine the role that metals play in the production of energy through the oxidation of carbohydrates (from our food). Oxygen, as the terminal electron acceptor in many of these processes, is sometimes converted into unwanted and dangerous by-products such as the radical anion superoxide, O_2^-. Several metal-containing enzymes remove these substances from the body. We shall consider the structure and action of these. The process of respiration produces carbon dioxide but, as you might expect, a build-up of gas in the body is something to be avoided. Carbonic anhydrase catalyses the conversion of CO_2 into HCO_3^- (and the reverse reaction). This enzyme, like LADH, which you met in the previous chapter, has zinc at its active centre.

Photosynthesis is the process by which carbon dioxide and water from the air are converted into carbohydrates. Although the process is confined to plants and bacteria, we depend on it ultimately as our source of carbon compounds for the production of energy and building of proteins, DNA and other molecules. Photosynthesis is a very complicated process and involves two very different metal centres that you will study in this chapter. The first is the more well-known chlorophyll, which contains a magnesium atom; the second is a cluster of four manganese ions and a calcium ion, which plays an important part in the overall process.

The final process you will study is nitrogen fixation, which is confined to bacteria but is nonetheless essential for all life. Nitrogen is needed to synthesise proteins, for example. Most nitrogen-fixing bacteria use a system containing two proteins to reduce molecular nitrogen, N_2, to ammonia. One of these proteins contains an active centre with Fe–S clusters, the other in most bacteria has a molybdenum-containing cluster.

The roles these systems play may seem quite diverse. We can however identify some themes that link some or all of the processes.

- Many of these proteins contain a transition metal at the active site and the function of the protein uses the ability of these ions to adopt several stable oxidation states.

- Many of the metals are in rigid environments. This reduces the amount of energy needed to rearrange atoms when a reaction takes place at the metal centre.

- The same structural motif reoccurs in a variety of processes. This is particularly true of the 4N-coordinating tetrapyrrole ring systems – porphyrins (which occur in haemoglobin, myoglobin, cytochromes c, cytochrome c oxidase, peroxidases and catalases), corrin (vitamin B_{12}) and chlorin (chlorophyll).

- The same proteins are used in different processes. For example, the electron transfer proteins you met in Chapter 7 are used not only in the processes we describe in this chapter but also in many other biological processes.

As you study this chapter, you should consider how these themes apply to each process and which of the roles outlined in Chapter 7 the metal play(s).

8.1   Oxygen transport in living systems

Air is essential for human life; more precisely, the oxygen in air is essential for life. As humans, we breathe in something like 45 000 litres of air every day simply to provide our bodies with a supply of oxygen.

Oxygen, O_2, comprises about 21% (by volume) of the air we breathe. It is transported via the bloodstream from the lungs to all parts of our bodies. The oxygen diffuses from the bloodstream into the cells, where it is used in aerobic respiration, the major process that provides energy. Using glucose as an energy source, the overall reaction for aerobic respiration is:

$$C_6H_{12}O_6 + 6O_2 = 6CO_2 + 6H_2O + \text{energy} \tag{8.1}$$

Six moles of oxygen are consumed for every mole of glucose, and a good supply of oxygen is essential to enable our cells, and ultimately our bodies, to operate and function normally. However, not only humans require oxygen; the life of most organisms, from the smallest single-cell amoeba to the largest elephant also depends on supplies of this simple molecule.

For small, single-cell organisms, oxygen is simply and easily obtained. These organisms make use of the facts that oxygen is slightly soluble in water and that it is a small molecule, which can quickly penetrate or diffuse through cell membranes.

As single cells are very small (ranging from 1 to 100 μm in diameter), the passive diffusion of oxygen into the centre of the cell is fast enough to support the respiration reactions. However, the amount of oxygen that can diffuse passively through the cell drops off rapidly with the distance over which the oxygen has diffused. What this means in practice is that organisms that rely on the passive diffusion of oxygen as their source of oxygen cannot be larger than about 1 mm in diameter; for larger organisms the oxygen would not get through in large enough quantities to support respiration.

Temperature is also important. The solubility of oxygen in water falls with increasing temperature. At 5 °C the solubility of oxygen in water is about 2 mmol dm^{-3}, which is enough oxygen in solution to maintain the respiration rate of a unicellular organism. Thus, very small organisms living at temperatures of about 5 °C are able to obtain their oxygen requirement by passive diffusion. However, at 40 °C the solubility is only around 1 mmol dm^{-3}.

Two immediate problems must be overcome to satisfy the oxygen requirements of large organisms (like humans), namely:

1 The rate of passive diffusion of oxygen through respiring tissue (e.g. skin) is not fast enough to penetrate much further than about 1 mm.

2 The solubility of oxygen drops off with increasing temperature.

The solubility of oxygen in blood plasma (the fluid component of blood, which does not contain red blood cells) at 37 °C is 0.3 mmol dm^{-3}. So, for warm-blooded organisms, like humans, the solubility of oxygen in blood plasma is not high enough to support aerobic respiration in the cells.

To survive, large animals (that is, greater than 1 mm in size) must have a means of capturing oxygen from the air, circulating it around the animal and, if they are warm-blooded or exist in hot climates, find a way of concentrating oxygen within their circulation systems.

The first problem of circulation is largely a mechanical one; a pump and pipes are required. Of course, these are the heart and blood vessels. The second problem of increasing the concentration of oxygen within circulation systems is largely a chemical one. It is this problem and the biochemical systems that overcome it, which will be the main subjects of this section.

As a final thought before we address the oxygen concentration problem in detail, consider the Antarctic ice-fish. This fish has a heart and circulation system similar to all vertebrates. However, it has no means of concentrating oxygen in its bloodstream (in fact, its blood is completely colourless). These fish live in temperatures of about −1 °C.

■ Why does the ice-fish have no biochemical oxygen concentration system?

□ At these temperatures the solubility of oxygen in water (or colourless blood) is higher even than at 5 °C, high enough to support respiration in the cells of the fish, so it has no need of a chemical system to concentrate oxygen in its bloodstream.

The solubility of oxygen in water at −1 °C is about 5 mmol dm^{-3}.

8.1.1 The chemistry of oxygen

To begin with, we need to examine the chemistry and properties of oxygen in some detail. O_2 is a very reactive molecule. Almost all the reactions of organic compounds with oxygen result in oxidation of the compound and reduction of oxygen to water. This type of reaction is nearly always highly exothermic. For example, fire is the oxidation by oxygen of organic matter (wood, oil, coal, etc.). The amount of energy given out from such a reaction is obvious.

The reason for the high reactivity of oxygen stems from its strongly oxidising nature. The reduction of oxygen to water is very energetically favourable. O_2 can actually be reduced not only to water but also to a variety of oxygen-containing molecules and radicals. Some of these are shown in Figure 8.1. These species can combine with H$^+$ and some of the resulting species are also

Metals and Life

shown. In this case the final species formed will depend on the pH of the solution, for example the fully reduced species can be OH^-, H_2O or H_3O^+.

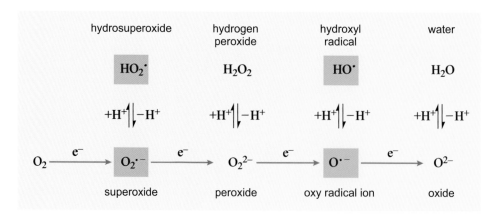

Figure 8.1   Species obtained on reduction of O_2. Radical species are printed on a darker background.

The properties of different oxidation states of the oxygen-containing molecule can be illustrated using a molecular orbital diagram.

■   Sketch the partial molecular orbital diagram for O_2 by combining 2p atomic orbitals on two oxygen atoms. Use the molecular orbital diagram to explain why:

O_2 has a double bond and O_2^{2-} has a single bond

the vibrational frequencies decrease in the order $O_2 > O_2^- > O_2^{2-}$.

☐   Figure 8.2a shows the partial molecular orbital diagram for O_2. Figure 8.2b shows the O_2 molecular energy levels for O_2, O_2^- and O_2^{2-}, each with a different number of valence electrons. As we move from O_2 to O_2^- (superoxide) to O_2^{2-} (peroxide), the π antibonding orbitals gain electrons. This gain results in a gradual weakening of the O_2 bond (decrease in bond order) and a consequent reduction in the vibrational frequency of the molecule. O_2^{2-} has a single bond compared to the double bond in O_2. It turns out that the vibrational frequency is a useful indicator of the formal charge on the O–O species. We shall return to this property when we come to study O–O molecules in living systems.

A closer inspection of Figure 8.2b shows that the superoxide anion, O_2^-, contains a single unpaired electron.

■   What properties does this confer on O_2^-?

☐   It is a radical, and is likely to be very reactive indeed. It is also paramagnetic.

Figure 8.1 also shows that, dependent on the pH, reduction of O_2 to water goes via a number of radical molecules; these are superoxide ($O_2^{\bullet-}$), hydrosuperoxide ($HO_2^{\bullet-}$), oxy radical anion ($O^{\bullet-}$) and the hydroxyl radical

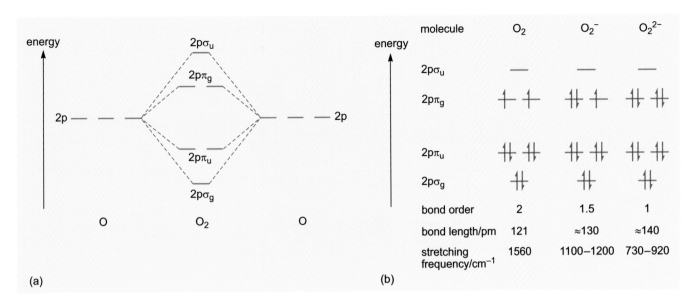

Figure 8.2 (a) Partial molecular orbital energy-level diagram for O_2 generated by combining 2p atomic orbitals on two oxygen atoms. (b) Molecular orbital energy-level diagrams for O_2, O_2^- and O_2^{2-}, showing orbital occupancies, and comparative data for the O–O bond in each of these species.

(`OH`). Further, hydrogen peroxide easily cleaves to form two hydroxyl radicals in the reaction

$$H_2O_2 \rightarrow 2 \ ^\bullet OH \tag{8.2}$$

These radicals have uncontrollably high reactivity and will react with almost any organic molecule. If these reactive species are formed in living systems (as we saw for $O_2^{\bullet -}$ in Box 3.2), they cause extensive and irreparable cellular damage by reacting with important biochemical molecules such as DNA, proteins and enzymes. Therefore, the 'handling' of oxygen in biochemical systems must be very carefully controlled so as to avoid the generation of detrimental radicals. We shall see later on in Section 8.2 that biochemical systems carefully manage the generation of oxygen radicals. In Section 8.3 we shall see how some enzymes are actually capable of utilising oxygen-containing radicals in biochemical reactions.

8.1.2 Transition metal–oxygen complexes

In view of the oxidising ability of O_2, it is not surprising that it reacts readily with transition metals. Transition metals participate in redox reactions. They exhibit a range of stable oxidation states and this property is a characteristic feature of transition metals; main-Group elements generally exhibit a narrower range of stable oxidation states. In fact, biological systems make use of this property of transition metals, and it is important for us to study briefly the types of metal–oxygen complexes that have been prepared.

For the purposes of this section, we shall focus our attention on metal–oxygen complexes that exhibit reversible binding of O_2. In other words, we are concerned here with those complexes that will bind O_2 under high pressures of O_2, and, in their oxygenated form, release O_2 at low pressures. The first complex that was found to exhibit reversible binding of O_2 was Vaska's complex, **8.1**. O_2 reacts readily with **Vaska's complex** to give compound **8.2**, which contains a peroxo ligand, coordinated to the iridium in a *side-on* position:

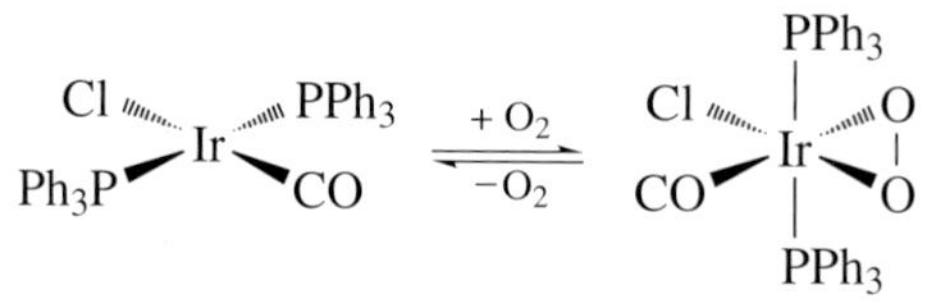

(8.3)

8.1   **8.2**

■   What are the formal oxidation states of the iridium ion in this reaction?

☐   In Vaska's complex, iridium is in a formal +1 oxidation state, because Cl^- is the only charged ligand. In the product of its reaction with O_2, the iridium is now formally in its +3 oxidation state because the oxygen has bonded as a peroxo ligand and is formally O_2^{2-} in a side-on position. (It is important to note here that the assignment of oxidation state to iridium is only a formal one. It is impossible to tell exactly what its oxidation state is in the O_2 complex. The reasons for this will become clearer when we examine the Fe–O_2 molecular orbital energy-level diagram later in this chapter. However, you should note that for the following examples of transition metal–O_2 complexes, it is difficult to assign the oxidation state of the metal unambiguously.)

Removal of O_2 from the oxygenated iridium(III) complex is achieved by reducing the amount of free O_2 in the reaction vessel. In accordance with Le Chatelier's principle, the original iridium(I) complex and free O_2 are regenerated.

Another example comes from cobalt chemistry (note that cobalt and iridium are in the same Group in the Periodic Table). Reaction of $[Co(NH_3)_6]^{2+}$, **8.3**, with O_2 results in the reversible substitution of one of the ammine ligands by O_2. In the product, **8.4**, the O_2 is bound through one oxygen atom in the mode known as end-on coordination:

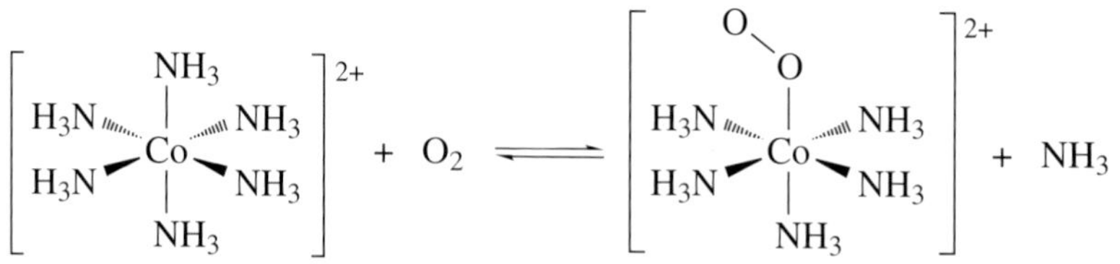

(8.4)

8.3   **8.4**

A slight variation on the cobalt reaction produces an interesting result. Increasing the amount of complex with respect to oxygen leads to another compound (Reaction 8.5). In this case the O_2 forms a bridging ligand between two cobalt pentammine groups to give the diamagnetic molecule **8.5**:

$$2 \left[\begin{array}{c} \text{NH}_3 \\ \text{H}_3\text{N} \text{\textbackslash} \text{Co} \text{\textbackslash} \text{NH}_3 \\ \text{H}_3\text{N} \text{Co} \text{NH}_3 \\ \text{NH}_3 \end{array} \right]^{2+} + \; O_2 \; \rightleftharpoons$$

8.3

$$\left[\begin{array}{c} \text{NH}_3 \quad \text{H}_3\text{N} \quad \text{NH}_3 \\ \text{H}_3\text{N} \text{Co} \text{O} \text{O} \text{Co} \text{NH}_3 \\ \text{H}_3\text{N} \quad \text{NH}_3 \quad \text{H}_3\text{N} \quad \text{NH}_3 \\ \text{H}_3\text{N} \end{array} \right]^{4+} + \; 2\text{NH}_3 \qquad\qquad (8.5)$$

8.5

As a final example of reversible O_2-binding to a transition metal, we shall consider a copper complex. Equation 8.6 shows the reaction of *tris*[3,5-di-isopropyl(pyrazolyl)]borate ethanenitrilecopper(I), **8.6**, with O_2. (Structure **8.6** is shown abbreviated as **8.7** in Equation 8.6.)

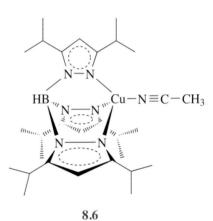

8.6

$$2 \; \begin{pmatrix} \text{N} \\ \text{N} \text{-Cu-N} \equiv \text{C-CH}_3 \\ \text{N} \end{pmatrix} \; \rightleftharpoons \; \begin{pmatrix} \text{N} & \text{O} & \text{N} \\ \text{N-Cu} & | & \text{Cu-N} \\ \text{N} & \text{O} & \text{N} \end{pmatrix} \qquad (8.6)$$

8.7   **8.8**

$+ \; O_2$ $+ \; 2\text{CH}_3\text{CN}$

The resulting O_2 complex, **8.8**, containing two copper atoms bridged by O_2 exhibits another mode of O_2-binding in which the oxygen coordinates in a bidentate manner to both metals.

From the examples above, it can be seen that oxygen will react in a reversible manner with transition-metal complexes to give a variety of oxygen coordination modes (summarised in Figure 8.3).

η^1	η^2	$\mu - \eta^1,\eta^1$	$\mu - \eta^2,\eta^2$
$O\!-\!O$ M	$O\!-\!O$ V M	M \ O / O \ M	M /\ O\!-\!O \/ M
8.9	**8.10**	**8.11**	**8.12**

Figure 8.3 O_2 coordination modes with transition-metal complexes.

■   Which properties of transition metals make them ideal for reversibly binding oxygen?

☐   (i) a range of stable oxidation states

(ii) a range of different coordination numbers.

We shall see that these properties are utilised in biochemical systems for oxygen transport.

8.1.3   Haemoglobin and myoglobin

From the discussion at the beginning of Section 8.1, it is clear that larger organisms must have a system for concentrating and circulating O_2 within their bodies; otherwise the passive diffusion of O_2 into the interior of the organism would be far too slow to support aerobic respiration reactions. From a chemical point of view, we also know that such organisms are likely to make use of transition metals in O_2 transport systems. The chemical properties of transition metals make them ideal centres for binding oxygen reversibly. We shall also see that another property of transition metals – the ability to form highly coloured complexes – is useful in characterising any transition metal-containing protein we study.

The brilliant red colour of blood comes directly from a chemical group called haem, which contains the transition metal iron. More specifically, the haem is found in the blood's O_2-carrying protein, haemoglobin (Hb). Haemoglobin is present in the bloodstream of many organisms.

Haemoglobin (Hb) is a medium-sized protein with a relative molecular mass of about 64 500. It is found in the red blood cells (erythrocytes) of many organisms, and its function is to concentrate oxygen in the bloodstream and transport it around the organism. Red blood cells are packed with Hb, which usually constitutes about 33% by mass of the cell. Whole blood contains about 120–180 g of Hb per dm^3. The solubility of oxygen in blood at 37 °C is about 10 mmol dm^{-3} (compare this with the solubility of oxygen at 37 °C in water, which is about 1 mmol dm^{-3} and in blood plasma, which is about 0.3 mmol dm^{-3}).

Myoglobin (Mb), on the other hand, is a smaller protein with a relative molecular mass of about 17 800. It is found exclusively in muscle tissue, where it acts as an oxygen storage site and also facilitates the transport of oxygen through muscle. As muscles require a lot of energy to operate, they also require very efficient access to oxygen in order to maximise respiration rates.

Both Hb and Mb have been the subject of intensive study. But what we know about both proteins comes almost entirely from a knowledge of their three-dimensional structures. The structures of myoglobin and haemoglobin were determined by J. C. Kendrew and M. F. Perutz, respectively, using what was then the relatively new technique of single-crystal X-ray diffraction. Hb can be regarded as a Mb tetramer (that is, approximately four Mb units linked together). Therefore, we shall study the structure of Mb in some detail, before considering Hb.

8.1.4 Structure of myoglobin

The higher-order structure of Mb is shown in Figure 8.4.

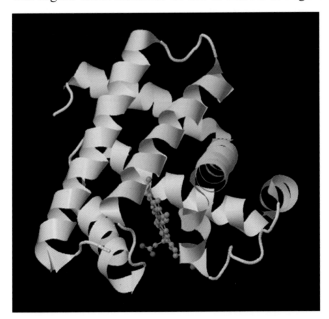

Figure 8.4 The higher-order structure of myoglobin, showing the large α-helical content of the protein. (Based on pdb file 2v1k (Hersleth et al., 2007).) The haem is shown in a ball and stick representation.

The protein consists mostly of α-helices of polypeptide. Sandwiched between two long α-helices lies a porphyrin ring. The porphyrin in Mb has various side chains attached to the outside of the ring. This particular derivative is known as protoporphyrin IX – you saw this structure in Figure 2.4d.

As you have seen previously, an iron(II) atom lies at the centre of the ring, coordinated by the four nitrogen atoms of the pyrrole groups. Note that the whole haem group (apart from the side chains) is planar (**8.13**). (The double bonds shown in **8.13** are for one resonance structure; in reality the porphyrin is extensively delocalised.) Mb, like Hb, has a strong red colour, which arises from $\pi–\pi^*$ transitions within the extensively delocalised haem group. On oxygenation of deoxy-Mb the colour of the protein changes from a dull red to a bright red (Figure 8.5). This colour change identifies the haem group as the site of O_2-binding within the protein.

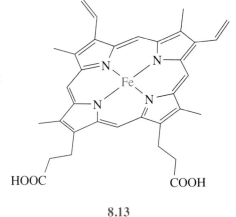

8.13

A closer examination of the structure shows that the haem group is attached to the protein backbone via a histidyl side chain. This coordinating histidyl side chain is known as the **proximal histidine**.

■ What is the approximate coordination geometry around the iron?

☐ The histidyl side chain and the porphyrin group give a square-pyramidal coordination geometry around the iron atom.

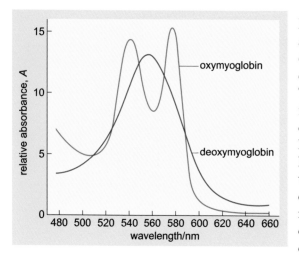

Figure 8.5 Visible spectra of oxymyoglobin and deoxymyoglobin. The bands are due to π–π* transitions within the haem group. The fact that the spectrum changes on oxygenation shows that the haem group is the O_2-binding site within the protein.

Figure 8.6a shows this coordination geometry in which the iron atom is 'out of the plane' of the porphyrin ring by about 60 pm. As iron(II) is usually six coordinate, the iron atom is coordinatively unsaturated in the deoxy form of the protein. The active site also contains another histidyl side chain, which is not coordinated to the iron atom. This histidyl side chain is on the opposite side of the haem group to the proximal histidine, and is known as the **distal histidine**; it turns out to have an important role, which we discuss below. The rest of the active site is taken up with hydrophobic amino acid side chains such as valine, which prevent possible dimerisation reactions between two Mb molecules in the presence of oxygen. (See the reaction of $[Co(NH_3)_6]^{2+}$ with oxygen described in Section 8.1.2 as an example of such a dimerisation reaction.)

From single-crystal X-ray diffraction studies on oxy-Mb, it was discovered that oxygen binds directly to the iron atom in an η^1-end-on fashion (Figure 8.6b). The further oxygen atom of the O_2 ligand participates in a hydrogen bond with the protonated nitrogen atom of the distal histidine. This extra interaction essentially stabilises the oxygen at the active site,

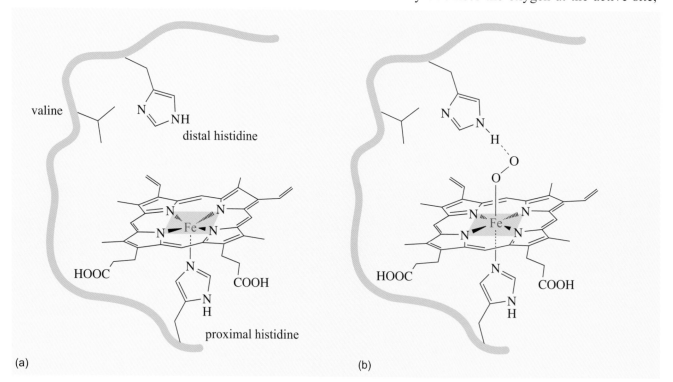

(a)   (b)

Figure 8.6 (a) Structure of the active site in deoxymyoglobin; showing the square-pyramidal iron coordination. (b) Structure of the active site of oxymyoglobin, showing the hydrogen bond from the distal histidine to the bound oxygen molecule; the distorted octahedral iron geometry can be seen at the centre.

causing it to bind preferentially over other small ligands. Binding of oxygen also results in a significant change in the iron coordination geometry. Remember that in the deoxy form, the iron lies 60 pm out of the basal plane towards the nitrogen atom of the proximal histidine. In oxy-Mb the iron is pulled into the plane of the porphyrin ring.

In effect the coordination geometry can now be described as distorted octahedral (Figure 8.6b).

8.1.5 Magnetic properties of myoglobin

O_2-binding not only causes a change in the structure and visible absorption spectrum of Mb (see Figure 8.5), but it also causes a change in the number of unpaired electrons in the haem group. Magnetic studies of metal ions in biosystems are often carried out and yield interesting insights into structures and mechanisms. It can, however, be difficult to interpret the data. This is not so for Mb. An accurate diamagnetic correction is known, which allowed reliable magnetic measurements to be performed. They showed that in the deoxy state, Mb is paramagnetic, with four unpaired electrons in the haem group, whereas in the oxy state, Mb is diamagnetic with no unpaired electrons. To understand this, we need to apply a little molecular orbital theory.

- ■ First, however, use crystal-field theory to sketch the d-orbital energy levels for a 3d transition metal in an octahedral crystal field. Show how the relative energies of the d orbitals change as the octahedral field is changed to a square-pyramidal field. Assume that the fourfold axis is the z-axis. Feed in the appropriate number of electrons for iron(II) in both diagrams, assuming a weak-field situation.

- □ The sketch in Figure 8.7 shows that on distortion of an octahedral field to a square-pyramidal one, the d_{z^2} orbital energy falls relative to that of $d_{x^2-y^2}$. Also the degeneracy of the d_{xy} orbital with the d_{xz} and d_{yz} orbitals is removed in that d_{xy} moves to a higher energy relative to d_{xz} and d_{yz}.

The four unpaired electrons for deoxy-Mb can be explained fairly easily by reference to the d-orbital energy-level diagram for iron(II) in square-pyramidal coordination geometry (Figure 8.7). Iron(II) has six d electrons; if we assume that it is in the high-spin configuration, then there will be four unpaired electrons.

To explain the diamagnetism of oxy-Mb, we need to examine the orbitals that interact with each other when O_2 binds to the iron atom. O_2 binds η^1 end-on to iron with an Fe–O–O angle of about 140°. This arrangement allows for a direct σ-overlap between the d_{z^2} orbital of the iron atom and the highest occupied orbital of O_2 ($2p\pi_g$) which lies in the iron xz-plane (Figure 8.8a). In addition the other $2p\pi_g$ orbital of the O_2 bonds weakly in a π-bonding fashion with the d_{yz} orbital of the iron (Figure 8.8b).

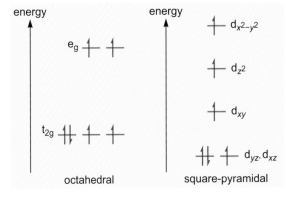

Figure 8.7 Crystal-field energy-level diagrams for high-spin octahedral and square-pyramidal iron(II).

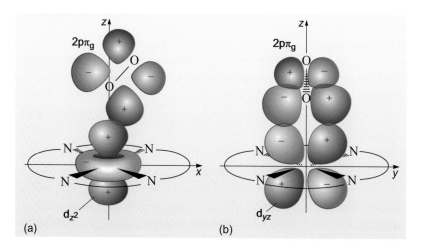

Figure 8.8 Overlap of O_2 and iron orbitals in oxymyoglobin. The $2p\pi_g$ orbitals of the oxygen molecule form (a) a σ bond with the d_{z^2} orbital on the iron, and (b) a weak π bond with the iron d_{yz} orbital.

We are now in a position to sketch a simplified molecular orbital energy-level diagram for the Fe–O_2 complex. The molecular orbital diagram in Figure 8.9 shows how the energy levels change when the d_{z^2} orbital of the iron atom interacts in a σ-bonding fashion with one of the $2p\pi_g$ orbitals on the oxygen molecule, and how the other $2p\pi_g$ orbital interacts with the d_{yz} iron atomic orbital. σ-bonding and antibonding orbitals are formed from the interaction of the iron d_{z^2} and one O_2 $2p\pi_g$ orbital. π-bonding and antibonding orbitals are formed from the iron d_{yz} and the other O_2 $2p\pi_g$ orbital. The iron d_{xz} and d_{xy} orbitals do not overlap to any great extent with either O_2 $2p\pi_g$ orbital, and remain non-bonding. We have eight electrons to put into Figure 8.8, six from iron(II) and two from the $2p\pi_g$ orbitals of O_2 (see Figure 8.2b).

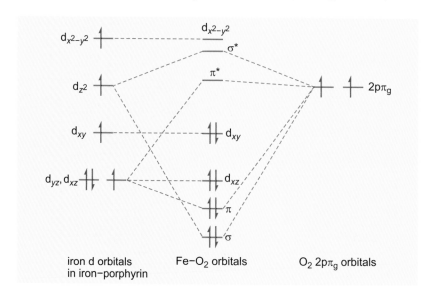

Figure 8.9 Molecular orbital energy-level diagram for the Fe–O_2 complex in oxymyoglobin.

Notice that the electrons are all paired, and hence the complex is diamagnetic.

Although it is customary to refer to deoxymyoglobin as containing Fe(II) and oxymyoglobin as containing Fe(III), this is one of a few cases where it is difficult to assign unambiguously the oxidation state of either the iron or the O_2, as the relative energies of the O_2 orbitals and the iron orbitals before formation of the Fe–O_2 complex are very similar. The corresponding molecular orbitals that are formed in the Fe–O_2 complex have roughly 'equal' amounts of iron and O_2 character. In other words, the electrons in the molecular orbitals are shared roughly equally between iron and O_2. It is therefore not strictly correct to describe it as an Fe(III)–O_2^- complex, although it can be said to contain some Fe(III)–O_2^- character, along with some Fe(II)–O_2 character.

There is one more point we should note about the arrangement of electrons in deoxy-Mb and oxy-Mb. In the deoxy state the iron(II) is clearly high spin. However, from the diamagnetism of the oxy state, the iron here must be low spin. Therefore, on O_2-binding, the iron goes from high spin to low spin.

8.1.6 Myoglobin and carbon monoxide

Before we leave Mb to concentrate on the intricacies of Hb, it is worth examining the role of the distal histidine in Mb. The distal histidine is essential for the successful function of Mb. Without it, the protein has a much lower affinity for oxygen and a much higher affinity for other small molecules. This shouldn't be too surprising. We have already seen in Figure 8.6b that the N–H group of the distal histidine forms a stable hydrogen bond with the further oxygen atom of the O_2 ligand, which increases the affinity of Mb for oxygen. Nevertheless, other small ligands will bind to the iron, essentially blocking the O_2-binding site. If this binding of small ligands is irreversible, the Mb ceases to function, with catastrophic consequences for the organism. Two examples of small ligands that are efficient at binding to iron are carbon monoxide and cyanide. Both substances are well known for their high toxicity to nearly all life forms. Indeed, both molecules are capable of irreversibly binding to Mb and blocking the access of O_2 to this binding site.

As the binding of both CO and CN^- to haem is so efficient, only very small concentrations of either are needed to block completely the O_2-binding site.

■ What is the distinctive feature of the bonding of CO (and CN^-) to transition metals that gives such strong bonding?

□ The binding stability of CO (and CN^-) to a transition metal is enhanced via π-acceptor back-bonding.

Fortunately, the concentration of CN^- in the air is too low to cause any problems. By comparison, the concentration of CO is relatively high, especially in built-up city areas. If gas fires have blocked flues, high concentrations of CO can build up, creating a poisoning hazard. CO is also present in vehicle exhausts and tobacco smoke.

To see how Mb prevents the catastrophic irreversible binding of CO to the haem site, we need to look carefully at the position of the distal histidine. Its position is such that it lies above the vacant binding site on the iron atom. As O_2 binds with an Fe–O–O angle of between 140° and 175°, the distal histidine does not sterically clash with the bound O_2 (Figure 8.10a).

■　At what angle does CO normally bond to a metal?

☐　CO ligands are bound linearly; that is, the Fe–C≡O angle is 180°.

In Mb, if CO does bind to the iron site, steric crowding by the distal histidine precludes the formation of a linear –Fe–C≡O complex (Figure 8.10b). In this way, CO binding to Mb is inhibited. This can be confirmed by comparing the CO-binding affinity of free haem with the CO-binding affinity of deoxy-Mb: the binding affinity of a free haem group for CO is about 25 000 times that for O_2 (CO is a much better ligand), whereas the binding affinity of Mb for CO is only about 250 times greater than that for O_2. In other words, CO is a very effective poison of the haem group, and, in the absence of the distal histidine, CO would bind so strongly to Mb that atmospheric O_2 could not displace it to any great extent. However, the presence of the distal histidine reduces the affinity of Mb for CO by a factor of 100 compared with free haem, and this is sufficient for the relatively high pressure of O_2 in the atmosphere to displace any CO bound to Mb. Remember that the CO partial pressure in air is relatively low, whereas that of O_2 is approximately 0.2 atm. Increasing the CO pressure in air eventually leads to effective competition of CO for the Mb-binding sites: this is when CO poisoning can occur.

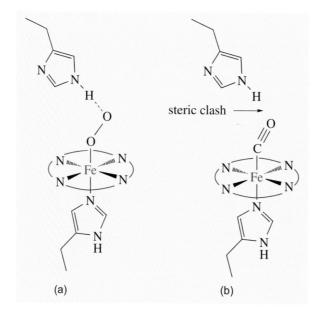

Figure 8.10　Active site structures of (a) oxymyoglobin and (b) carboxymyoglobin. In carboxymyoglobin the carbon monoxide sterically clashes with the distal histidine. (The Fe–C≡O angle is a matter of debate, but is thought to be in the range 140–175°.)

8.1.7 Haemoglobin

At first sight the structure of Hb appears to be very complex (see Figure 1.9).

However, it is easier to consider Hb as essentially four Mb units linked together. Therefore, the Hb molecule has four haem units and, hence, four O_2-binding sites. The active sites of the haem units in Hb are almost identical to the active site structure in Mb, complete with proximal and distal histidines. As a result, we might expect to see exactly the same O_2-binding properties in Hb and Mb. If this were so, it would cause a problem for the organism. Remembering that Mb appears in muscle and Hb appears in the bloodstream, we can see that Hb must deliver O_2 to Mb. If both Hb and Mb had precisely the same O_2-binding characteristics, the transfer of O_2 from Hb to Mb would be inefficient, certainly not efficient enough to maintain an appropriate O_2 concentration in the muscle tissue. However, if we look at Figure 8.11, we can see that Mb *does* bind O_2 more efficiently than Hb at low partial pressures of O_2 (similar to the O_2 partial pressures found in a working muscle) and therefore oxygen transfer from haemoglobin to myoglobin does proceed.

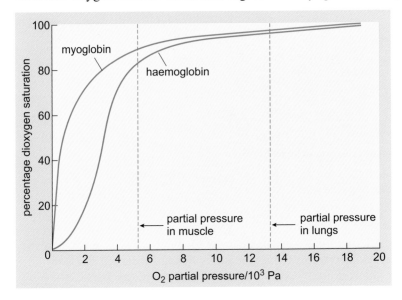

Figure 8.11   Percentage saturation of haem sites as a function of O_2 partial pressure for myoglobin and haemoglobin. Myoglobin follows a hyperbolic curve, whereas haemoglobin follows a sigmoidal curve due to a positive cooperative effect between the individual haem sites.

What property of Hb causes this difference in O_2-binding efficiency, despite the structural similarity of the O_2-binding sites? To answer this question, we need to look at the structure of Hb in more detail.

■   When O_2 bonds to Mb, how does the position of the iron change?

☐   The iron moves into the plane of the porphyrin ring.

As in the Mb case, O_2 binding to the haem site in Hb causes a movement of the iron atom into the plane of the ring (Figure 8.6). From very extensive structural studies of Hb, it turns out that this motion of the iron atom on O_2-binding causes changes in the rest of the protein's higher-order structure. These structural changes alter the O_2 affinity of the other haem sites in Hb. In other words the four haem sites in Hb can 'communicate' with one another, the communication being in the form of a structural change. This type of communication within a protein is known as an **allosteric interaction**. In Hb, the binding of O_2 to one haem site increases the O_2 affinity of the remaining haem sites; this is known as the **cooperative effect**. (As the binding of the first O_2 increases O_2 affinity at the remaining sites, this is called a positive cooperative effect.)

From a structural point of view, the changes in Hb on progressive O_2-binding are extremely subtle and still not fully understood. However, several key features have been observed in the crystal structures of the deoxygenated and the oxygenated forms of Hb. It appears that there are two distinct forms of Hb; these are known as the tense form (abbreviated to **T-state**) and the relaxed form (abbreviated to **R-state**). The structures of the T-state and R-state are shown in Figure 8.12.

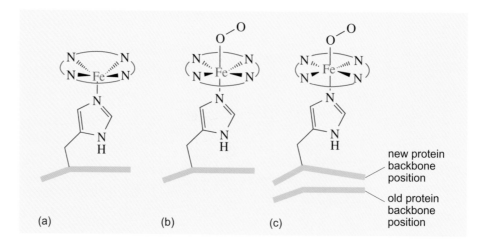

Figure 8.12   Haem group geometry changes on O_2-binding in haemoglobin; the haem group is abbreviated to four nitrogen atoms linked by curved lines: (a) deoxy active site showing iron atom out of porphyrin plane; (b) T-state, showing O_2 bound to iron atom, but iron atom is not fully in plane of porphyrin ring; (c) R-state, showing iron in plane of porphyrin ring and the motion of the protein backbone responding to the motion of the iron atom.

In completely deoxygenated Hb, the iron atom is out of the plane of the porphyrin ring in all the haem binding sites (Figure 8.12a). On the binding of one O_2 atom to the first haem site, the iron atom moves slightly towards the plane of the porphyrin ring, but not fully into the plane (Figure 8.12b). The reasons for this are fairly complex, but the proximal histidine is the root cause. It prevents the complete movement of the iron(III) into the plane of the porphyrin ring because it is anchored to the protein. The most stable iron(III) coordination geometry, for the singly oxygenated protein, is when the iron is in the plane of the porphyrin ring. But as the iron is slightly out of the plane of the porphyrin ring, the system is under tension – hence the term 'T-state'. At some point, probably on binding of further O_2 molecules to other haem sites, the tension becomes too great for the protein structure. At this point the iron(III) atoms move completely into the plane of the porphyrin rings (Figure 8.12c). This motion causes a significant change in the position of the proximal histidine, which, in turn, changes the position of the protein backbone; the protein structure is now in its relaxed state or R-state. But this is not the end of the protein movements. The change in the position of the protein backbone is mechanically transmitted to all of the haem sites. In the R-state, O_2 binding to deoxyhaem gives the iron(III)–O_2 complex in which the iron atom is directly in the plane of the haem. Because the tension of the protein is released on transition to the R-state (after O_2-binding), this process has been called the **Perutz trigger mechanism** ('trigger' owing to the nature of the process).

Perhaps the simplest way of depicting the trigger mechanism is shown in Figure 8.13. The two forms of the protein are shown. The R-state is favoured when Hb is oxygenated, the T-state is favoured on deoxygenation.

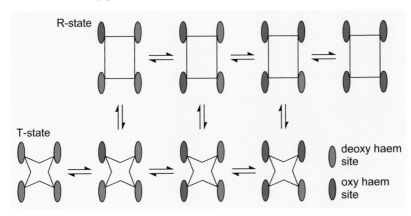

Figure 8.13 Schematic diagram of T- and R-states of Hb: the higher the degree of oxygenation, the more likely the R-state is to dominate.

How do these subtle structural changes affect the binding of O_2 to Hb? At low oxygen partial pressures, Mb can remove O_2 from Hb. At high oxygen partial pressures (when Hb must pick up O_2 as efficiently as possible – that is, in the lungs), the R-state dominates and the affinity of Hb for O_2 is relatively high. Figure 8.11 shows these results in graphical form. The O_2-binding curve for Hb has a sigmoidal shape. In contrast, O_2-binding affinity to Mb follows a hyperbolic shape. A hyperbolic curve shows that the percentage of Mb molecules that are oxygenated depends simply on the oxygen partial pressure. For the sigmoidal curve, the percentage of Hb molecules that are oxygenated depends not only on the oxygen partial pressure, but also on the state of Hb.

The Hb story does not end here. Despite all the advances that have been made in understanding how Hb works, the protein is still the centre of much research attention. There are questions that remain partially or completely unanswered, but recent research has provided an answer for others. For example, how does a fetus obtain oxygen from its mother's Hb? See Box 8.1.

Box 8.1   Fetal haemoglobin

So that the transfer of oxygen can be effected, fetal Hb must have a higher affinity for oxygen than maternal Hb. In adult haemoglobin, HbA, the four myoglobin subunits are of two types α and β with some differences in the amino acid residues. The tetramer can be described as $\alpha_2\beta_2$. Oxygen release from oxyhaemoglobin is regulated by 2,3-diphosphoglycerate (2,3-DPG), which binds to the interface of the two β-units and promotes the allosteric shift from the R- to the T-state. In fetal haemoglobin, HbF, however, the two β-subunits are replaced by two γ-subunits. Again the γ-subunit differs from the α- and β-subunits in some of the amino acid residues; in particular a serine is substituted for a histidine that is involved in the binding of 2,3-DPG. This removal of two positive charges from the binding site (one contributed by each β-chain in HbA) reduces the affinity of 2,3-DPG for HbF, making the oxygen-binding affinity of fetal haemoglobin higher than that of the mother – therefore oxygen can be transferred from maternal to fetal blood cells.

8.1.8   The role of the higher order structure of myoglobin

The structures and chemistry of both Mb and Hb are complex and yet fascinating. The haem groups and protein structure are linked perfectly to give the optimum O_2-binding, transport and storage system. Indeed, this is clearly seen if one takes a simple haem group (in the absence of any protein) and observes its O_2-binding properties. It is a well-known chemical fact that a free haem group reacts *irreversibly* with O_2 and could never act as an O_2-carrier on its own: this is because it reacts with oxygen in such a way that the O_2 molecule is eventually dissociated. The general reaction scheme for a model compound (but with a pyridine in place of histidine) is shown in Figure 8.14. Two haem groups (Figure 8.14a) react with O_2 to give a μ-peroxo dimer (Figure 8.14b); this is analogous to the μ-peroxo dimer shown in Reaction 8.5.

■   Why does this dimerisation not take place in Mb or Hb?

▫   The bonded O_2 is stabilised by hydrogen-bonding to the distal histidine. The steric requirements of the proximal and distal histidines and the protein backbone in Mb or Hb prevent the approach of a second haem group.

This dimer then dissociates to give the reactive oxoiron(IV) haem (Figure 8.14c), which reacts with further haem groups to give μ-oxoiron(III) haem dimers (Figure 8.14d) The overall result is that the oxygen molecule has been irreversibly split.

For quite some time, it seemed impossible that an isolated haem group could recreate the reversible O_2-binding seen in Hb/Mb. However, in 1978, in a piece of simple but highly innovative thinking, James Collman hit on the idea

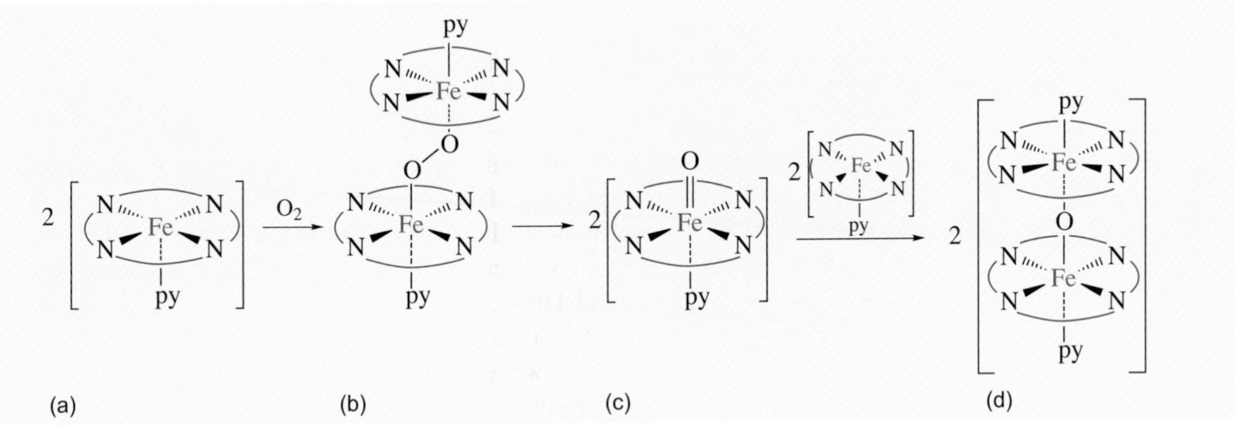

Figure 8.14 (a) Reaction of free haem (with pyridine; py) group with O_2. In the first instance, a μ-peroxo haem dimer (b) is formed, which then cleaves to give an oxoiron(IV) haem complex (c). The oxo complex quickly reacts with further haem groups to give μ-oxoiron(III) haem dimers (d). The result is that the O_2 molecule has been irreversibly cleaved.

that if the dimerisation of haem in the presence of O_2 could be prevented, then reversible O_2-binding might be observed (Burke et al., 1978).

■ How might the dimerisation of the haem be prevented?

☐ As in the natural systems, it might be possible sterically to prevent the approach of a second haem by putting other groups in the way.

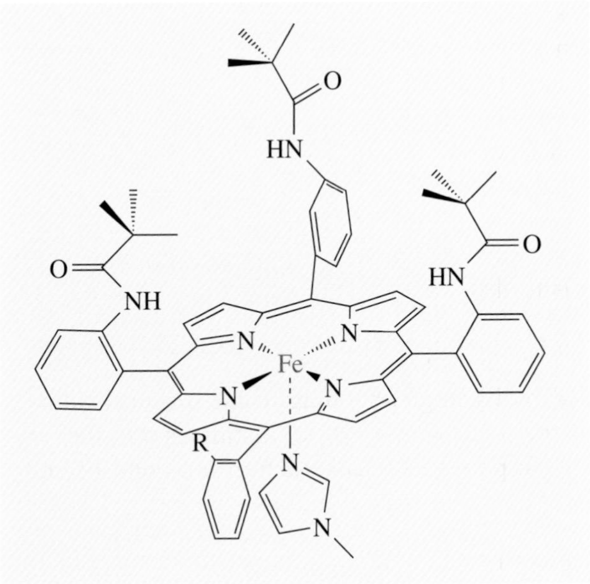

Figure 8.15 Collman's iron picket-fence porphyrin, showing a haem group, to which four aromatic groups are attached (between two adjacent pyrrole rings); the benzene rings have bulky side chains in *ortho* positions (for clarity, one has been represented as 'R', and the haem side chains have been omitted). The side chains form a sterically crowded 'fence' around the site where O_2 binds to iron. This complex exhibits reversible O_2-binding, similar to that exhibited by myoglobin and haemoglobin.

Collman prepared the haem group derivative shown in Figure 8.15. He called the molecule an **iron picket-fence porphyrin**. The haem group has four bulky groups attached to it. These groups all stick up around the O_2-binding site like a fence. This fence prevents two haem groups from coming together (but does not hinder the binding of O_2), and thus precludes the formation of μ-peroxy haem dimers following the binding of O_2. Indeed, Collman found that the iron picket-fence porphyrin could reversibly bind O_2 in an analogous fashion to the haem groups in Mb, and that the O_2 affinity of iron picket-fence porphyrin was very similar to that of Mb.

The structures of Mb (Figure 8.4) and Hb (Figure 1.9) show that here too the haem groups are buried within the structures of the proteins; as with Collman's iron picket-fence porphyrin, this prevents dimerisation of the haem groups in the presence of oxygen. The burying of the oxygen-binding site within the protein also serves another purpose. The bound oxygen could be released as a superoxide radical, which, if it were liberated into the cell, could irreversibly damage other biomolecules. (See Box 3.2.) By keeping the oxygen in a protective pocket, both Mb and Hb prevent this detrimental reaction. The other O_2-carrying proteins we shall study also prevent the reaction of bound O_2 by keeping it in a deeply buried protein pocket.

8.1.9   Haemerythrin

Some animals have different coloured blood from the brilliant red colour of blood that contains Hb. These different colours are qualitative evidence that the O_2-carrying protein in these animals is not Hb. Some invertebrates have blood that is colourless in the absence of oxygen yet a deep blue colour in the presence of oxygen (true blue blood); the blood of certain marine worms is colourless in the absence of oxygen and deep burgundy in the presence of oxygen. However, the fact that all these species have coloured blood suggests that their oxygen transport systems could involve transition metal-containing proteins. This should not be surprising because many transition metals other than iron can exist in a variety of oxidation states, and this is the key property for metals in O_2-carrying proteins. In this section, we shall consider the O_2-carrying protein, **haemerythrin (Hr)**. It will be useful to compare and contrast the properties of this protein with what we already know about Hb and Mb.

Haemerythrin (Hr) occurs in the bloodstream of certain marine worms. It is a small protein with a relative molecular mass of about 13 500. Its higher-order structure has similarities to Hb (which we can consider to be a Mb tetramer), in that Hr is often found as an octameric unit, with a total relative molecular mass of about $8 \times 13\ 500 = 108\ 000$. At its active site, Hr contains two iron atoms. However, despite the 'haem' in its name, the protein does not contain a haem group. From single-crystal X-ray diffraction studies, the structure of the active site in the deoxy form has been determined to be **8.14**. The first point to make about the active site structure of deoxy-Hr is that it is completely different from that of Hb and Mb. Each iron atom is coordinated either by two or three histidine amino acid side chains (indicated by 'N' in **8.14**); they are also bridged by the carboxylate groups of glutamate and aspartate amino acid side chains and a hydroxide group between the two iron sites.

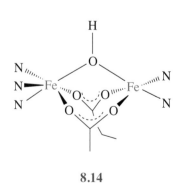

8.14

■ What is the difference in coordination between the two iron atoms?

□ One of the iron atoms is six coordinate in a roughly octahedral geometry. Its coordination sphere is made up of the nitrogens of three histidyl side chains and the oxygens of bridging glutamyl and aspartyl side chains and a hydroxide anion. The second iron atom is five coordinate with the same three bridging ligands but only two histidyl ligands.

Despite the structural differences between Hb/Mb and Hr, there are a couple of very important similarities.

■ Why is it important that one of the irons is five coordinate? Where does this occur in Hb and Mb?

□ A five-coordinate iron atom is coordinatively unsaturated. In other words this iron atom can bind a further ligand. The deoxy forms of Hb and Mb contain five-coordinate iron in the centre of the porphyrin ring.

The second similarity is that the iron atoms in the deoxy form are both in the +2 oxidation state, again similar to Hb and Mb. Our knowledge of how O_2 binds to the active site of deoxy-Hr comes from single-crystal X-ray diffraction of several Hr derivatives and from vibrational studies.

■ What experimental technique can be used to obtain the vibrational spectrum of O_2?

□ We could not use infrared spectroscopy to study the vibrations of O_2 because its dipole moment does not change during vibration. Raman spectroscopy can be used instead.

8.1.10 Resonance Raman studies of haemerythrin

The resonance Raman spectroscopy studies for the oxidised form of Hr, oxy-Hr, proved to be particularly revealing (Kurtz et al., 1976). They used ^{18}O-labelled O_2 as the oxygen source to obtain additional information.

■ What is the most abundant isotope of oxygen?

□ ^{16}O; this occurs naturally in 99.76% abundance, whereas the natural abundance of ^{18}O is 0.20%.

The labelled O_2 gas contains not just the normal ^{16}O isotope but contains an increased proportion of ^{18}O atoms.

Figure 8.16 depicts the O_2 part of the resonance Raman spectrum of oxy-Hr oxygenated with 58% ^{18}O-labelled O_2. (50%-labelled gas would contain a statistical mixture of isomers, 25% ^{16}O–^{16}O, 50% ^{18}O–^{16}O and 25% ^{18}O–^{18}O. 58%-labelled gas would contain less ^{16}O–^{16}O and more ^{18}O–^{18}O.)

■ What is immediately striking about the stretching frequencies in Figure 8.16?

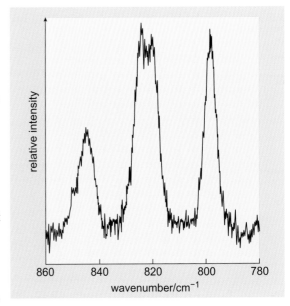

Figure 8.16 Resonance Raman spectrum of oxyhaemerythrin oxygenated with 58% ^{18}O-labelled O_2. The wavelength of the excitation radiation used was 514.5 nm.

□   The Raman frequencies due to the O–O bond are now much lower than in free molecular O_2 (Figure 8.2b).

■   What oxygen species would give rise to these frequencies?

□   A glance at Figure 8.2b shows that these frequencies correspond to a peroxide. In other words, the oxygen must have formally gained two electrons from the protein to give O_2^{2-}.

You may also have noticed that there are now four bands in the spectrum, corresponding to four vibrations. For free ^{18}O-labelled O_2, we would expect to see three vibrations corresponding to ^{16}O–^{16}O, ^{18}O–^{18}O and ^{16}O–^{18}O. The two flanking bands correspond to ^{16}O–^{16}O and ^{18}O–^{18}O vibrations. (With equal proportions of Hr–^{16}O–^{16}O and Hr–^{18}O–^{18}O, we would expect absorptions of similar peak height. The height of the absorption at about 845 cm^{-1} is lower than the other flanking band because the isomeric ratio was not 1:1.) The central band due to the ^{16}O–^{18}O vibration can be considered as two vibrations.

How can we account for the central ^{16}O–^{18}O vibration splitting into two bands? There are two possible ways of bonding ^{16}O–^{18}O to a single five-coordinate iron atom: these are with either the ^{16}O bonded directly to the iron (structure **8.15**) or the ^{18}O bonded directly to the iron (structure **8.16**). These two structures do have slightly different ^{16}O–^{18}O stretching frequencies, so we would expect to see four O–O vibrations in the resonance Raman spectrum.

What are our conclusions from the resonance Raman experiments? The number of bands in the resonance Raman spectrum strongly suggests an η^1 coordination of O_2 to the active site. The vibrational frequency of the bound O_2 in the spectrum of oxy-Hr is very low and shows that it is bound as the peroxide form, so it must have received two electrons from the active site.

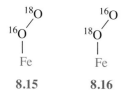

8.15   **8.16**

■   Which iron atom do we expect the O_2 to bond to?

□   Remembering the structure of the deoxy form (**8.14**), it seems reasonable that in oxy-Hr, O_2 binds in an η^1 fashion to the five-coordinate iron atom.

■   The evidence shows that the oxygen binds formally as O_2^{2-}. What will have happened to the oxidation state of the iron atoms?

□   Both iron atoms are formally oxidised to iron(III), but remember that it is difficult to assign the oxidation state unambiguously.

This process is shown in Figure 8.17, where the structure of the active site is indicated in both deoxy-Hr and oxy-Hr. The structure of the oxy-Hr active site has been inferred from single-crystal X-ray diffraction studies on various Hr derivatives. These crystal structures predict that the O_2 binds as a hydroperoxide (taking a hydrogen atom from the bridging hydroxide group).

■   Despite the structural differences between Hr and Hb/Mb, there are several common features. Try to list them now.

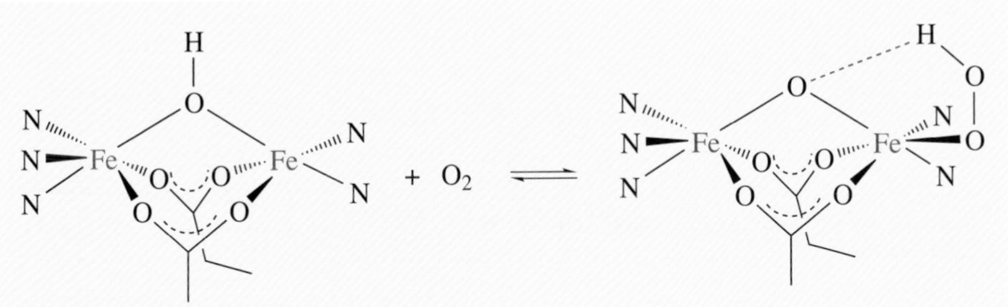

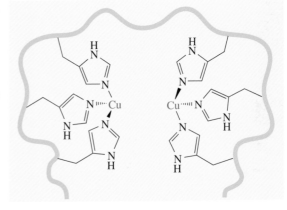

8.14

Figure 8.17 Binding of O_2 to haemerythrin and structure of the active sites. Note that there is an apparent transfer of hydrogen from the bridging hydroxide to the bound oxygen molecule; this is thought to help stabilise the binding of O_2 similar to the role of the distal histidine in myoglobin and haemoglobin. The resultant hydroperoxide is hydrogen-bonded to the oxide bridge.

◻ The reduction of O_2 to a lower oxidation state by iron; the formal oxidation of iron(II) to iron(III); η^1-binding of O_2 to a five-coordinate iron atom; and the stabilisation of the bound O_2 with a hydrogen bond.

8.1.11 Haemocyanin

The third, and last, type of O_2-carrying protein we shall study is called **haemocyanin (Hc)**. This protein occurs in the bloodstreams of molluscs (e.g. snails) and arthropods (e.g. spiders). It is a very large protein with a relative molecular mass of around 1 000 000. In the deoxygenated state it is colourless, whereas in the presence of O_2 it is blue. At its active site, Hc has two copper atoms. In fact, the name 'haemocyanin' is a double misnomer, as the protein contains neither a porphyrin ring nor an iron atom, although it is blue as reflected in 'cyan'. The structure of the active site is shown in Figure 8.18.

Structurally, the active site in Hc is similar to that in Hr, with two metal ions available for O_2-binding. The copper atoms in the deoxy form are in the +1 oxidation state.

Figure 8.18 Active site of deoxyhaemocyanin. The copper-to-copper distance is about 360 pm (for scale, each copper-to-nitrogen bond distance is around 190 pm).

■ What is the electronic configuration of a copper(I) ion? Would you expect it to exhibit d–d transitions in a UV–visible spectrum?

◻ Copper(I) is d^{10}. This helps explain why the deoxy protein is colourless; as the d orbitals are full, there cannot be any d–d electronic transitions.

Copper(I) normally adopts a tetrahedral geometry; hence it has a coordination number of four. In this active site, each copper is coordinated by three nitrogen atoms of histidyl side chains. As in the previous cases of deoxy-Hb, deoxy-Mb and deoxy-Hr, the metal atoms are coordinatively unsaturated.

■ What is different about this active site from the other three that we have studied?

☐ In deoxy-Hc, both metal atoms are coordinatively unsaturated, so that both atoms can accept another ligand.

How does O_2 bind to the active site in the oxygenated form? The first piece of evidence in answering this question came, once again, from resonance Raman studies (Thamann et al., 1977). In an exactly analogous manner to the experiment carried out with Hr, 50% ^{18}O-labelled O_2 was used in the resonance Raman experiments. The results are shown in Figure 8.19.

■ What are the two key features that you should look for in this spectrum?

☐ (i) The position of the peak gives information on the strength of the O–O bond and therefore the extent of its reduction.

(ii) The number of peaks gives us a clue about the mode of bonding to the metals.

First, we note that the O–O vibrational frequency is about 730 cm^{-1}.

■ What does this frequency tell us about the charge on the bound O_2?

☐ The characteristic wavenumber of the peroxide ion O_2^{2-} is 730–920 cm^{-1} and so O_2 is bound as peroxide. The charge on the bound O_2 will be −2.

■ What can we say about the formal oxidation state of the copper in the oxygenated form?

☐ Both copper ions have been formally oxidised from oxidation state +1 to oxidation state +2.

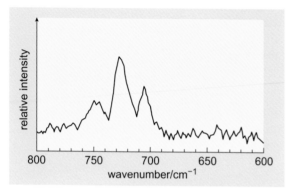

Figure 8.19 Resonance Raman spectrum of oxyhaemocyanin oxygenated with 50% ^{18}O-labelled O_2.

Second, the spectrum contains only three peaks. In contrast to Hr, the resonance Raman spectrum of Hc shows no broadening or splitting of the ^{16}O–^{18}O vibration. Hence the O_2 must be bound *symmetrically* at the active site. Reference to Figure 8.3 shows that we can rule out a structure of type **8.9** as a possible O_2-binding geometry. We can also rule out structure **8.10**, as it is very likely that the O_2 will bind to both copper ions simultaneously. But how can we tell whether it has a structure of type **8.11** or type **8.12** (Figure 8.20)?

The question of the O_2-binding geometry in Hc was answered by a group of Japanese chemists, led by N. Kitajima (Kitajima et al., 1988). They had prepared the copper(I) compound **8.6**, the abbreviated form of which is shown on the left of Reaction 8.6 (repeated below). Notice that the ligand, *tris*[3,5-di-isopropyl(pyrazoyl)borate] mimics the three histidyl side chains coordinated to copper in Hc. The Japanese chemists noticed that their compound reversibly bound O_2 at low temperatures; the colour changes they

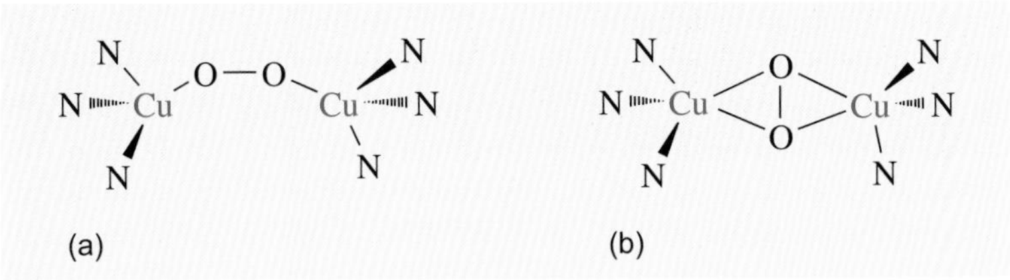

Figure 8.20 Two possible O_2-binding geometries in oxyhaemocyanin compatible with resonance Raman evidence. In both cases the O_2 is bound symmetrically between the two copper atoms.

saw matched exactly the colour changes seen on reversible oxygenation of Hc. They were able to obtain a single-crystal X-ray diffraction structure of the oxygenated product of their reaction; its structure **8.8** is shown as the product of Reaction 8.6. Notice that on oxidation a dimer has formed.

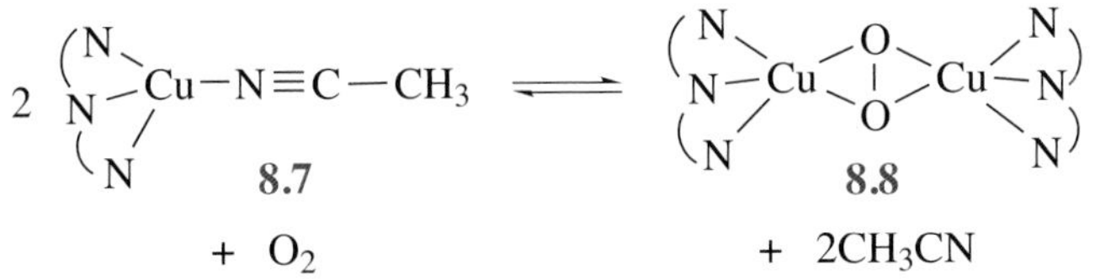

(8.6)

■ What is the nomenclature for the type of bridging formed by the O_2 here?

□ Oxygen has bound between the two coppers in a $\mu-\eta^2$, η^2 fashion.

Remarkably, the UV–visible spectrum of the oxygenated compound **8.8** matched almost exactly the UV–visible spectrum of oxy-Hc (Figure 8.21). This was good evidence that in oxy-Hc, O_2 binds between the two coppers as a $\mu-\eta^2$, η^2 peroxide. Indeed, this was confirmed by a single-crystal X-ray diffraction study of oxy-Hc in 1994 (Magnus et al., 1994).

8.1.12 Some final thoughts on oxygen carriers

Box 8.2 Green blood

Green blood featured in an episode of CSI (The Theory of Everything, the fifteenth episode in the eighth season of the popular American crime drama *CSI: Crime Scene Investigation*). Green blood may sound like science fiction and indeed Star Trek's Mr Spock supposedly had green blood, but real cases have been noted of patients whose blood is a dark green.

The cause of the colour is the replacement of haemoglobin by sulfhaemoglobin. In sulfhaemoglobin, there is substitution of sulfur into the porphyrin ring. Spectroscopic evidence has been recorded that suggests that the porphyrin group has been converted into a chlorin ring containing sulfur. The uptake of oxygen is reduced in patients whose blood contains sulfhaemoglobin as it binds O_2 less strongly. Sulfur can be incorporated into haemoglobin from sulfur-containing drugs. In a case

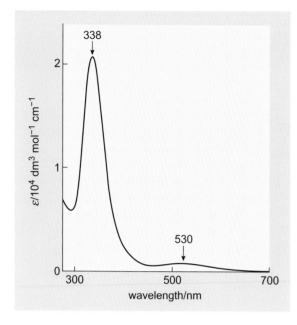

Figure 8.21 UV–visible spectrum of compound **8.8**. with bands at 530 nm ($\varepsilon =$ 840 dm^3 mol^{-1} cm^{-1}) and 338 nm ($\varepsilon =$ 20 800 dm^3 mol^{-1} cm^{-1}). The UV–visible spectrum of oxyhaemocyanin has two similar bands at 570 nm (ε approximately 1000 dm^3 mol^{-1} cm^{-1}) and 345 nm (ε approximately 20 000 dm^3 mol^{-1} cm^{-1}).

in Canada in 2007 (Flexman et al., 2007), as in the *CSI* episode, the patient had been taking large doses of a sulfur-containing migraine drug.

Having seen how reversible O_2-binding is successfully carried out in proteins, another question comes to mind; how do proteins/enzymes irreversibly bind O_2? After all, the proteins we have discussed in Section 8.1 merely transport O_2 to where it is consumed. There must be some proteins/enzymes that are capable of taking the delivered O_2 and using it to oxidise molecule(s) irreversibly. Take respiration, for example, which was discussed at the beginning of Section 8.1; how is the oxygen consumed in this process, and how is it converted into H_2O without releasing all the intermediate toxic molecules shown in Figure 8.1? In the next section, we shall look at some of the enzymes that irreversibly fix O_2 for use in a range of biochemical processes including respiration.

8.2 O_2 in respiration and other oxidation processes

We have seen in Section 8.1 that haemoglobin transports O_2 around the body, and that O_2 is bound by myoglobin in muscle tissue. But where is the O_2 delivered to and how does it react? These are important questions to ask, as we know that O_2 is a very reactive molecule, capable of forming potentially harmful radicals (see Section 8.1, Figure 8.1). To avoid the production of these radicals, the reactions of O_2 within the body must be very carefully controlled. Not surprisingly, this control is exerted by several enzymes.

In this section, we shall briefly look at what is known about enzymes that catalyse reactions involving O_2, superoxide (O_2^{2-}) and hydrogen peroxide (H_2O_2). These include the reaction of O_2 in aerobic respiration and the reaction of O_2 with biochemical molecules such as hormones, lipids and amino acids. We shall also study two classes of enzymes that catalyse reactions of H_2O_2 and one that destroys the superoxide ion. All these enzymes contain metals at their active sites. Finally we consider an enzyme, carbonic anhydrase, which deals with the gaseous product of respiration, CO_2.

8.2.1 O_2 in aerobic respiration – cytochrome c oxidase

We know from the equation for the aerobic respiration of glucose

$$C_6H_{12}O_6(aq) + 6O_2(g) = 6CO_2(g) + 6H_2O(l) \qquad (8.7)$$

that oxygen must be consumed within our bodies in the respiration process. Obviously our bodies do not consume oxygen as in a flame, as this would liberate uncontrollable amounts of heat. Oxygen consumption in living systems is a very carefully controlled reaction, which has been referred to as cold combustion. This occurs in series of stages; in this section we are concerned with the final stage which reduces O_2 to H_2O.

This takes place at the inner membrane of mitochondria. As you saw in Chapter 4, mitochondria are found within cells (Figure 4.1a showed a sketch of a typical animal cell). The mitochondria contain a heavily folded inner membrane which increases the available surface area. Oxygen is bound and reduced by one particular protein that is found embedded in the membrane. The protein is called **cytochrome c oxidase** (CCO). (You met this previously in Section 7.4.)

Cytochrome c oxidase (CCO) is an enzyme, which, as its name implies, oxidises cytochrome c as part of its function. It is a very large and complicated protein (relative molecular mass > 100 000). Like many proteins located in cell membranes, it has both hydrophilic and hydrophobic surfaces; it is therefore difficult to crystallise from either water or organic solvents. The determination of the X-ray crystallographic structure of bovine (Tsukihara et al., 1995) and bacterial (Iwata et al., 1995) CCO in 1995 was a major breakthrough. Chemical analysis had suggested that CCO consists of 13 subunits. The X-ray structure of crystals of the oxidised enzyme indicates a dimer with all 13 subunits present in each monomer. CCO contains a variety of metal active sites, which are all important in the protein's mechanism of action. X-ray studies of bovine heart CCO showed the presence of two iron ions (both surrounded by haem groups), three copper ions, a magnesium ion and a zinc ion. Figure 8.22 shows a sketch of the positions of the metal ions, other than magnesium, in CCO.

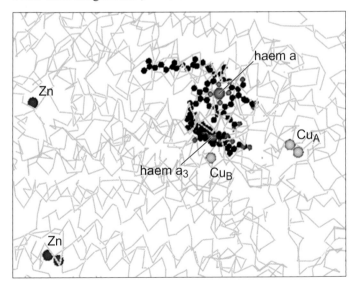

Figure 8.22 Sketch of a close up of CCO showing the positions of the metals. (Based on the pdb file 2eij (Muramoto et al., 2007).)

We shall concentrate on one site in particular, haem a_3, the site where O_2 is bound and reduced to water in the following reaction:

$$O_2(g) + 4H^+(aq) + 4e^- = 2H_2O(l) \tag{8.8}$$

But first we will consider how the electrons are delivered to the active site.

8.2.2   Mechanism of CCO

Cytochrome c delivers electrons to the Cu_2 group (Cu_A in Figure 8.22) near the outside of the protein. There are possible electron transfer paths from here to both the haem groups. However there is evidence that transfer is faster for the haem group, haem a, that is not the binding site for O_2. The two haem groups are relatively close and so electrons could reach haem a_3 via haem a. The O_2-binding site contains two metal atoms, iron in haem a_3 and copper. A coordination bond between a histidyl side chain and the iron atom (see haemoglobin and myoglobin) anchors the haem group to the rest of the protein. The copper atom is coordinated by three histidyl side chains, and is on the opposite side of the haem group to the histidine (Figure 8.23). The active site itself is reminiscent of the active sites in both myoglobin and haemocyanin.

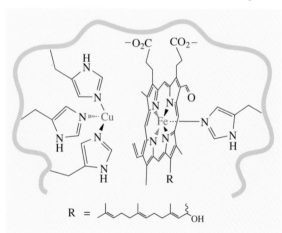

R = [structure] OH

Figure 8.23   Active site for O_2-binding in CCO.

The proposed mechanism for reduction of oxygen by CCO can be written in the form of a catalytic cycle. This summarises the addition of O_2 to CCO, its reduction to H_2O and regeneration of CCO ready to start again. Figure 8.24 shows this cycle. Starting with the reduced form of CCO (**8.17**), the reactant O_2 is added. This attaches to the iron of the haem group to form the intermediate **8.18**. Protons and electrons are then transferred to the active site from other parts of the enzyme resulting in cleavage of the O–O bond and production of the Fe(IV)–O intermediate, **8.20**, and water bound to Cu. Finally additional protons and electrons are transferred to remove the O ligand as water and reduce Fe and Cu back to their initial oxidation states (+2 and +1, respectively), recovering **8.17**.

In the reduced, active form of CCO (Figure 8.24, **8.17**), the O_2-binding site contains low-spin iron(II) and copper(I). The structure of fully oxidised CCO shows O_2 bound to iron and copper as a peroxide (μ–η^1, η^1), in which the metals are formally present as high-spin iron(III) and copper(II) (structure **8.18**, Figure 8.24).

However, a resonance Raman study carried out on reduced CCO in aqueous solution at 3 °C suggests that the mechanism in Figure 8.24 may not be correct. The researcher (Kitagawa, 2000) monitored the spectrum over time as O_2 attaches to the active site and then is reduced to H_2O. The wavelength of the incident laser light matched that of a transition of the Fe–haem group, so that it was groups attached to Fe rather than Cu that were monitored. In Section 8.1.10, the region of the spectrum corresponding to O–O stretches was used to help determine the method of attachment of O_2 to Fe. In the

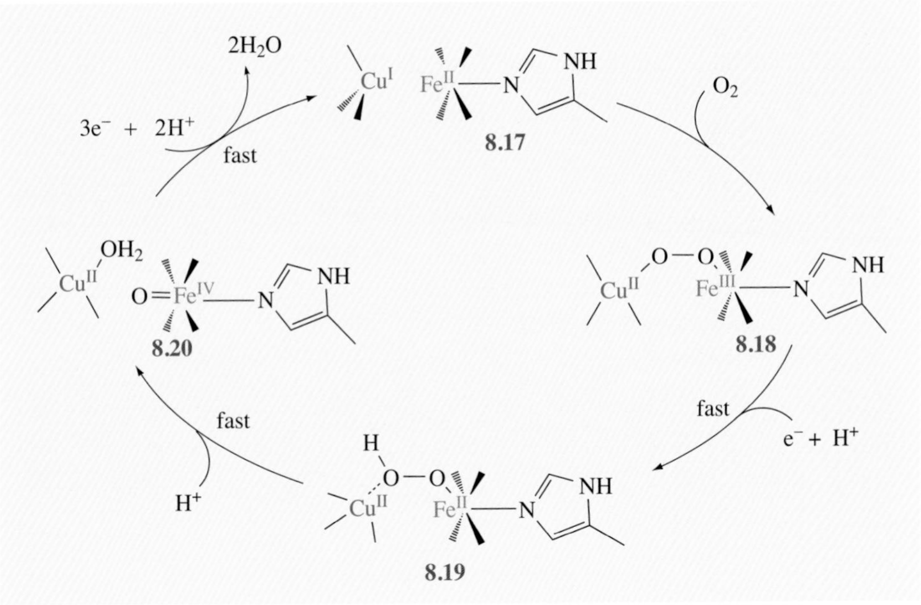

Figure 8.24 Catalytic cycle for the reduction of O_2 by cytochrome c oxidase (CCO).

experiment on CCO, vibrations at lower wavenumbers were monitored corresponding to Fe–O stretches. After 0.1 ms, one band from Fe–^{18}O–^{18}O, one from Fe–^{16}O–^{16}O and one each from Fe–^{16}O–^{18}O and Fe–^{18}O–^{16}O were observed.

■ What can you conclude about the binding of O_2 from the observation of separate bands for Fe–^{16}O–^{18}O and Fe–^{18}O–^{16}O?

☐ The O_2 is attached to the Fe in an η^1 fashion.

Following this, a new pair of bands corresponding to Fe–O stretches was observed. These bands occurred at the same wavenumbers as when either pure $^{16}O_2$ or when pure $^{18}O_2$ were used and no vibrations at intermediate wavenumbers were present.

■ What did this indicate about how the oxygen was attached to Fe?

☐ This indicated that only a single oxygen was attached to the iron. If O_2 were attached, we would expect to see bands due to $^{16}O^{18}O$.

These bands were assigned to Fe(IV)=O stretches. After 2.7 ms, these bands disappeared and bands assigned to Fe(III)–OH were recorded.

So on this timescale, it appears that O_2 attaches to Fe in an η^1 fashion. The O–O bond is then broken to give an Fe=O species. Finally this is protonated. There is no evidence for the fully oxidised bridging peroxide form of the enzyme, **8.18**. A further point that suggests the peroxide species of the fully oxidised state is not involved in the reduction of O_2 is that the reduction proceeds easily and rapidly in solution. The fact that the fully oxidised form can be crystallised and its X-ray diffraction pattern recorded implies it is stable and not a highly reactive reaction intermediate. Thus there is still work to do on determining the mechanism.

Most of the O_2 that is used for respiration is completely reduced to H_2O. Despite the high efficiency of CCO some partially reduced O_2 molecules do, however, escape into the rest of the cell. These potentially harmful molecules must be destroyed by the cell. This is accomplished by enzymes that act as detoxifying agents. Two particular classes of enzymes, **peroxidases** and **catalases**, specifically decompose H_2O_2. One other class, called superoxide dismutases, catalyses the decomposition of the superoxide radical anion, O_2^-. We shall examine examples of these enzyme classes later in this chapter.

Under extreme conditions of exercise, it is possible that the supply of electrons to CCO is too slow to prevent the release of significant quantities of partially reduced oxygen compounds. When this happens, cell damage can occur. Such a release of toxic molecules is known as a respiratory burst.

One more point about CCO is of interest. Cyanide is very effective at binding to haem groups, and it is likely that cyanide binds irreversibly to the haem group in CCO. As the amount of CCO is very small compared to that of other haem containing proteins like myoglobin and haemoglobin, CCO is very susceptible to cyanide poisoning. In fact, it is thought that the toxicity of cyanide is due to its ability to inhibit CCO function. Blockage of CCO by cyanide stops respiration immediately and the cell dies, leading quickly to organism death.

8.2.3   Cytochrome P450

So far we have looked at proteins that bind O_2 for use in respiration; this is the major use of O_2 in the body. But O_2 is also used as the terminal electron acceptor in a variety of oxidation reactions carried out by enzymes. One particular class of these enzymes is known as the **cytochrome P450s** (when carbon monoxide is added to these particular enzymes, a carbon monoxide complex is formed, which has a strong absorption in the visible region at 450 nm; hence the name P450). They catalyse the following reaction (where R is an alkyl group in a biomolecule):

$$R–H(aq) + O_2(g) + 2H^+(aq) + 2e^- = R–OH(aq) + H_2O(l) \tag{8.9}$$

What is remarkable about this reaction is that alkyl groups are rather inert. Certainly, reaction of common alkanes (e.g. propane, hexane) with air at room temperature is very slow, if not imperceptible. The cytochrome P450s (abbreviated to P450s hereafter) are powerful catalysts, capable of catalysing Reaction 8.9, and related oxidation reactions, at body temperature in aqueous solution and at relatively low partial pressures of O_2. Moreover, this reaction is carried out by the enzyme stereospecifically; in other words only one enantiomer is produced when the product is chiral. Some examples of the reactions catalysed by P450s are shown in Reactions 8.10–8.13:

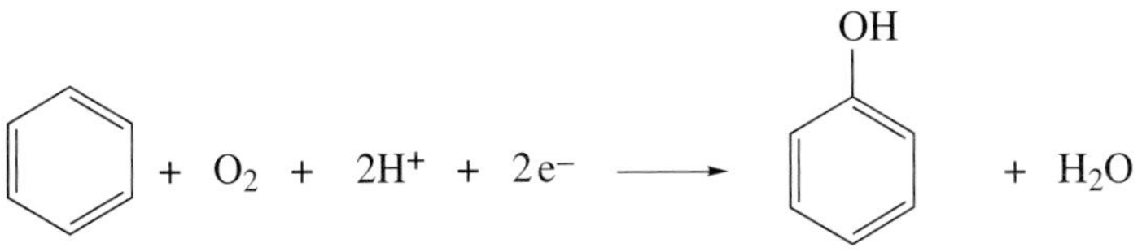

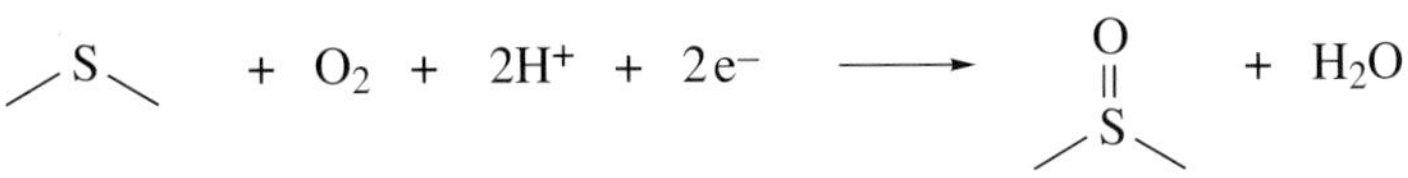

Such reactions are useful for converting unwanted molecules (known as xenobiotics) including drugs, carcinogens and pesticides into forms that can be removed from the body and for the synthesis of steroids from cholesterol. Unfortunately they can also convert some xenobiotics into carcinogens. As Reactions 8.10–8.13 show, the O_2 molecule is split in the reaction, with one oxygen atom ending up in a water molecule and the other in the product. As only one of the oxygen atoms ends up in the product, P450s are known as **mono-oxygenases**. (Other enzymes also use O_2 as a substrate, and in these cases both oxygen atoms end up in the product. Such enzymes are called 'di-oxygenases'; we shall not cover di-oxygenases in any detail in this book.)

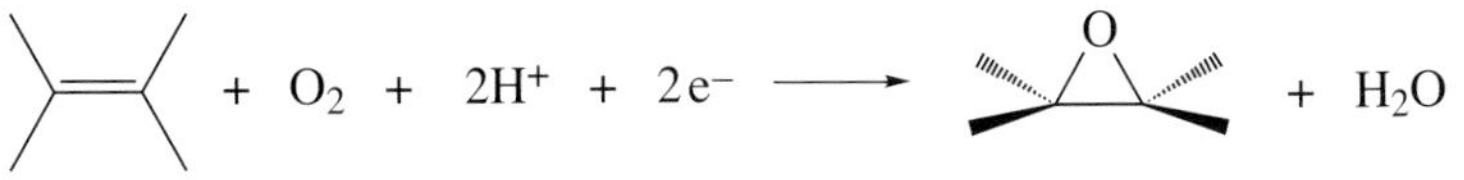

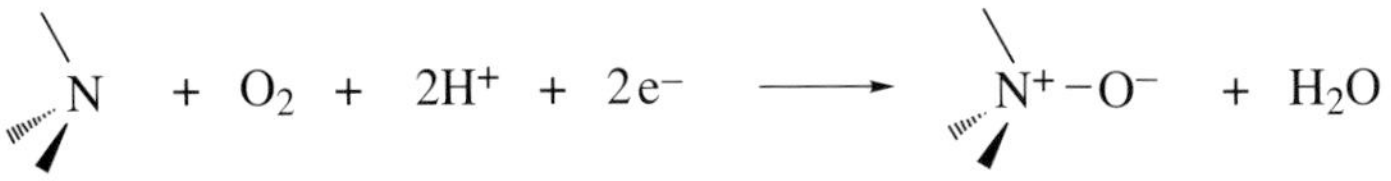

Several individual types of P450 exist, but all are thought to have the same basic higher-order protein structure. One example, camphor-5-mono-oxygenase has been crystallised and its crystal structure solved. The structure of part of the active site is shown in Figure 8.25. S represents a cysteinyl residue.

■ How would you describe the structure of the active site?

☐ The active site contains a haem group, the iron atom of which is also coordinated by the sulfur atom of a cysteinyl side chain.

■ Which group have we seen coordinating in this position in other proteins?

☐ In other proteins, we have only seen a histidyl side chain in this position.

On the opposite side of the haem to the sulfur atom is a cavity, which is formed by a rigid arrangement of hydrophobic amino acids (such as valine and leucine). This is an ideal binding site for hydrophobic substrates, like fatty acids.

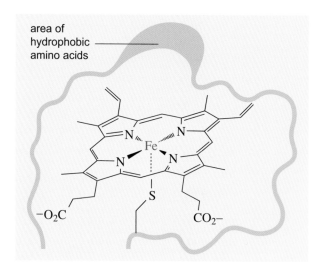

area of
hydrophobic
amino acids

Figure 8.25   A sketch of the active site of camphor-5-mono-oxygenase, a cytochrome P450, showing the hydrophobic pocket where the organic substrate binds; the structure of the iron coordination geometry at the active site is shown.

8.2.4   Catalytic cycle for cytochrome P450s

The proposed catalytic cycle for P450s is shown in Figure 8.26 (although, as in the last example, there is still considerable debate about whether this cycle is correct). In its resting state (structure **8.21**, Figure 8.26), the iron is in its low-spin +3 oxidation state and the enzyme is inactive. Attachment of the substrate converts this into a high-spin Fe(III) species. Electrons transfer from the outside of the protein to the active site, reducing iron(III) to iron(II). (The electron is obtained from NADPH acting as an electron carrier. The electron is then transferred from the surface of the protein through an electron transfer pathway, made up of amino acids, to the active site.)

In the iron(II) state (structure **8.22**, Figure 8.26), the iron atom can bind an O_2 molecule (structure **8.23**, Figure 8.26). This is quickly followed by addition of a proton and a further electron to give a hydroperoxide ligand coordinated to iron(III) (structure **8.24**, Figure 8.26). A further proton then reacts with the hydroperoxide to give a water molecule and a highly reactive porphyrin–iron(V)–oxo group (structure **8.25**, Figure 8.26). This group is formally Fe(V)–O although there is spectroscopic evidence that it is in fact an Fe(IV)–O group and the extra electron is removed from the surrounding ligands.

In contrast to CCO, it is proposed that in P450s the highly reactive porphyrin–iron(V)–oxo reacts with an organic molecule (denoted as RH in Figure 8.26), which is already bound in the active site adjacent to the haem group. Here we see the role of the hydrophobic cavity: the cavity binds the organic substrate, and, in so doing, brings the organic substrate in close proximity to the iron(V)–oxo group. The consequence is that when the iron(V)–oxo group is generated, it can react quickly with the organic substrate before the oxygen can be reduced to water with further electrons and protons. The last step of the cycle produces the monoxygenated substrate, ROH, and regenerates the enzyme in its resting state (structure **8.21**, Figure 8.26).

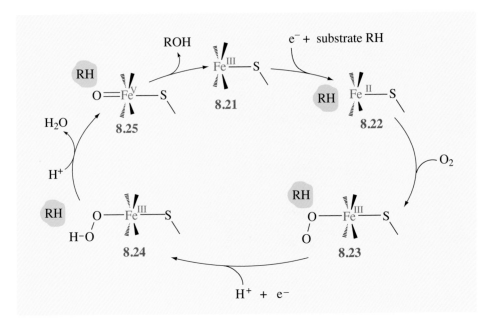

Figure 8.26 Catalytic cycle for the oxidation of organic molecules by cytochrome P450: structure **8.21**, resting state of enzyme; structure **8.22**, reduced state of enzyme before O_2-binding; structure **8.23**, after O_2-binding; structure **8.24**, hydroperoxide form; structure **8.25**, formation of reactive iron(V)–oxo species.

■ What other proteins will bind O_2 when they are in the iron(II) oxidation state?

☐ Haemoglobin, myoglobin, haemerythrin and CCO all contain iron(II) in their deoxygenated form. On oxygenation the iron(II) is formally oxidised to iron(III).

Note that P450s cannot bind O_2 in the iron(III) state and require an electron before becoming active. In the same way, if myoglobin or haemoglobin are oxidised to iron(III) (not by O_2) to give a deoxygenated iron(III) form, then this form cannot bind oxygen. Cooking red meat is an example of this. Fresh, uncooked meat is bright red. The colour is due to oxygenated haemoglobin and myoglobin. On standing (or cooking) the meat turns brown. The brown colour is the inactive iron(III) form of the proteins, with a water molecule coordinated to the iron of the haem group instead of O_2.

Turning back to the catalytic cycle of CCO, you will see parallels between the mechanisms of CCO and P450. The most important similarity is the proposed generation of a highly reactive iron–oxo species, either iron(IV)–oxo or iron(V)–oxo. It is the high reactivity of this species that allows both enzymes to react very quickly and efficiently.

8.2.5 Peroxidases and catalases

As you saw with CCO, respiration reduces O_2 usually to water, but sometimes to partially reduced O_2 molecules, like hydrogen peroxide. To address this potential problem, there are several classes of enzyme that are known to catalyse the destruction of H_2O_2. H_2O_2 is destroyed in the following reactions (where XH_2 is an organic molecule):

$$2H_2O_2 = 2H_2O + O_2 \tag{8.14}$$

$$H_2O_2 + XH_2 = X + 2H_2O \tag{8.15}$$

Reaction 8.14, which is the disproportionation of H_2O_2, is catalysed by catalases. The sole function of this class of enzyme is to remove H_2O_2. Reaction 8.15 is catalysed by peroxidases; in this reaction, H_2O_2 acts as a substrate to oxidise an organic molecule (in biochemical systems, these organic molecules can be fatty acids, phenols, amines, etc.). For example, a quinol with two OH groups will be oxidised to a quinone with two =O groups.

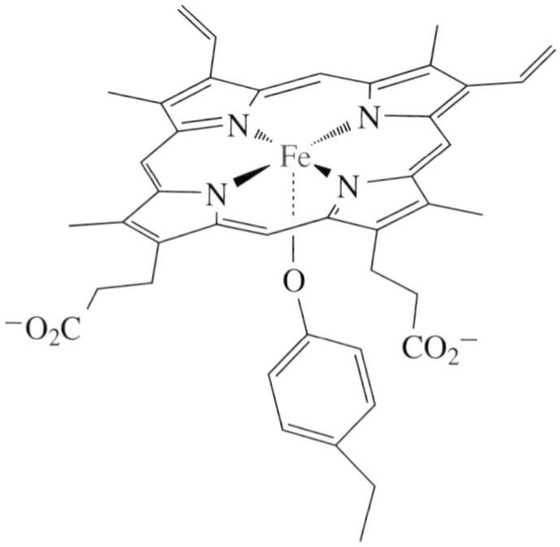

8.26

Typically, catalases and peroxidases have a haem group at their active sites. Structure **8.26** shows an example of a catalase in which the iron atom of a haem group is coordinated by a tyrosinate (tyrosine anion) side chain (in contrast to the histidyl and cysteinyl side chains that you have seen in this position in other proteins). Unlike in deoxygenated myoglobin, for example, here the iron is in its +3 oxidation state, the negatively charged tyrosinate ligand helping to lower the $(Fe^{3+}|Fe^{2+})$ couple relative to that in myoglobin.

The active site of peroxidases is very similar and also contains iron in the +3 oxidation state. Intermediates in the peroxidase catalytic cycle have been studied by a variety of techniques and the following mechanism inferred from the results:

$$[Fe(III)\ R] + H_2O_2 \longrightarrow \underset{\text{(compound I)}}{[Fe(IV)\!=\!O\ R^\bullet]} + H_2O \qquad \text{(step 1)}$$

$$\underset{\text{(compound I)}}{[Fe(IV)\!=\!O\ R^\bullet]} + XH_2 \longrightarrow \underset{\text{(compound II)}}{[Fe(IV)\!=\!O\ R]} + {}^\bullet XH + H^+ \qquad \text{(step 2)}$$

$$[\text{Fe(IV)}=\text{O R}] + \text{HX}^{\bullet} + \text{H}^{+} \longrightarrow [\text{Fe(III) R}] + \text{X} + \text{H}_2\text{O} \qquad (\text{step 3})$$
(compound II)

R represents the porphyrin group or an amino acid residue in the enzyme. In step 1, an oxygen atom attaches to the Fe(III) which is oxidised to Fe(IV) and an electron is removed from R to form a radical. The oxidised enzyme is green and is known as compound I. Because the enzyme has undergone a two-electron oxidation, compound I is sometimes considered an Fe(V) species, although there is spectroscopic evidence that the second electron is not localised on the Fe but is on the surrounding group R. In step 2, a substrate molecule XH_2 donates an electron to compound I producing the red Fe(IV)=O species (compound II) and forming a radical $^{\bullet}XH$. Finally oxygen is removed from the Fe to re-form the original state of the enzyme and $^{\bullet}XH$ is oxidised to X.

Catalases form compound I in a step similar to step 1 above but do not go on to form compound II. Instead they form a peroxide complex with a hydrogen peroxide molecule hydrogen bonded to the oxygen of the Fe(IV)=O group. The hydrogen peroxide then reduces the Fe(IV)=O group directly to Fe(III) producing H_2O and O_2 in the process.

8.2.6   Cu–Zn superoxide dismutase

Copper–zinc superoxide dismutase (abbreviated to Cu–Zn SOD hereafter) catalyses the following reaction:

$$2O_2^{-} + 2H^{+} = O_2 + H_2O_2 \qquad (8.16)$$

In Section 8.1.1 (Figure 8.1), we saw that the O_2^{-} ion (superoxide ion) is a radical molecule (see also Figure 8.2b). It is extremely reactive and will indiscriminately react with cell components. Even trace quantities of superoxide, which might come from incomplete reduction of O_2 by CCO, are hazardous. Cu–Zn SOD is a very efficient enzyme, which catalyses the reaction of superoxide to form the less harmful O_2 and H_2O_2.

■   How will a system then deal with hydrogen peroxide?

□ H_2O_2 is later removed by catalases or peroxidases.

The active site (see Figure 2.3a), does not contain a haem group and the Cu–Zn SOD's mechanism of action does not involve an iron–oxo species.

A proposed mechanism is shown in Figure 8.27 in which the copper is the active metal, changing between oxidation states +1 and +2 during the catalytic cycle. In the first few steps, superoxide binds to the copper (displacing water). Copper(II) is then reduced to copper(I) with the concomitant release of O_2. The bridging imidazolate is also protonated in these steps (structures **8.27–8.29**, Figure 8.27). In the next step, another superoxide molecule and a proton enter the active site. The superoxide coordinates to the copper(I) and also hydrogen bonds to two protons (structure **8.30**, Figure 8.27). The copper(I) reduces the superoxide to peroxide, which is then released as hydrogen peroxide, and is itself reoxidised to copper(II). An alternative mechanism for reduction of O_2^- to H_2O_2 involves the O_2^- ion gaining a proton from the bridging imidazole. The OOH$^-$ ion then either attaches to Cu, forming a Cu(II) complex, **8.31**, or obtains an extra proton from a nearby water molecule to form H_2O_2 without ever complexing with Cu. The final step in both cases is loss of H_2O_2 and formation of the initial Cu–Zn SOD, **8.27**.

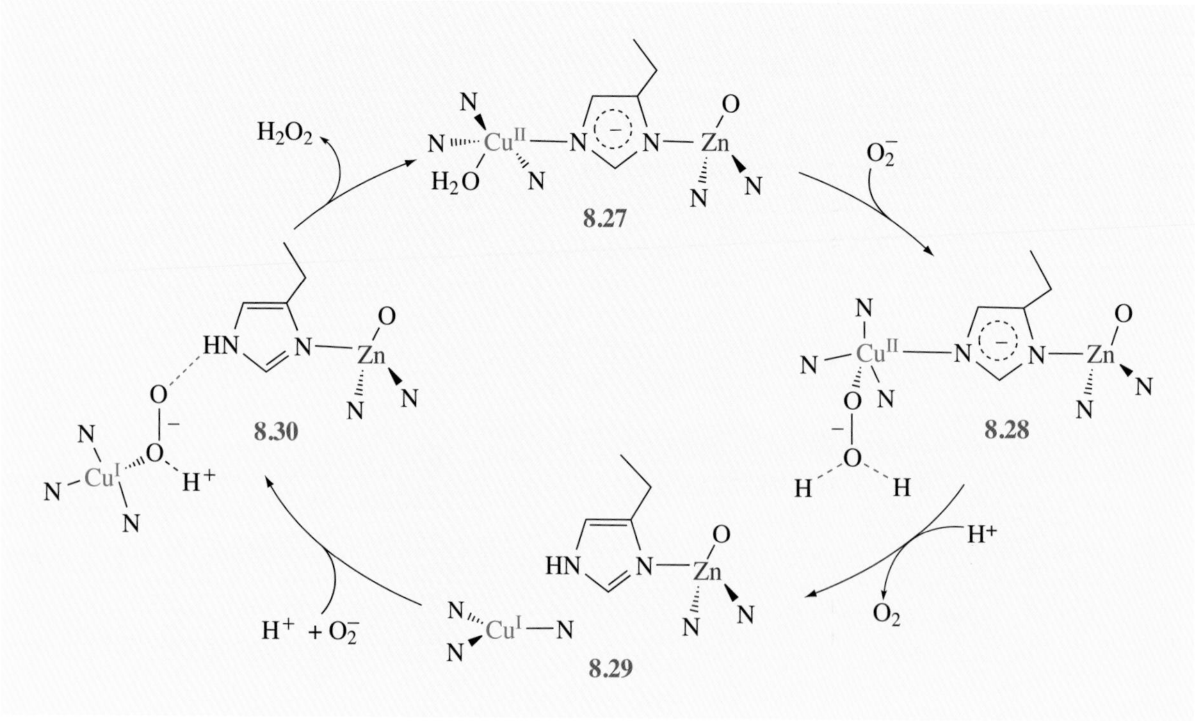

Figure 8.27 Catalytic cycle for the destruction of superoxide by copper–zinc superoxide dismutase. Structure **8.30** shows the second superoxide substrate coordinating to the copper and forming two hydrogen bonds (dashed lines). In structures **8.27**–**8.30** coordinating histidyl groups are represented by 'N' and the aspartate group by 'O'.

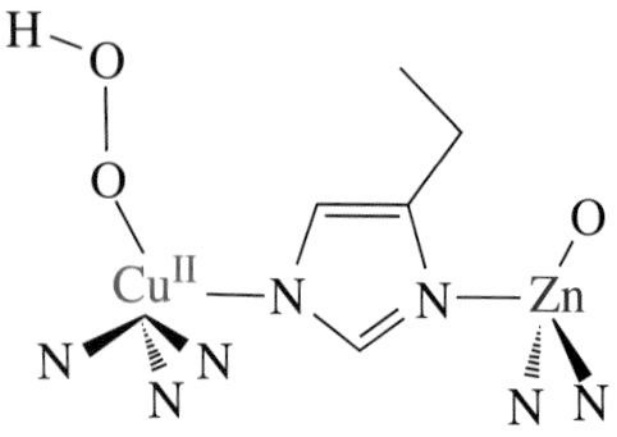

8.31

Cu–Zn SOD is not the only superoxide dismutase. For example, Mn superoxide dismutase is found in human mitochondria and, as indicated in Chapter 3, plants need manganese to produce an Mn SOD.

■ Why is Mn a suitable metal for the active site of a SOD?

☐ Mn has stable oxidation states differing by one (Mn(II), Mn(III) and Mn (IV)) and can therefore form a complex that will undergo one-electron reduction.

The oxidised state of Mn SOD contains Mn(III).

8.2.7 Carbonic anhydrase

Carbonic anhydrase (CA) is an enzyme that is present in many organisms, including humans, where it is found in the erythrocytes (red blood cells). It was first discovered in 1932, and in 1939 it was shown that CA contained zinc. The enzyme catalyses the following physiologically important reaction, where carbon dioxide and water react to make hydrogen carbonate, HCO_3^- (commonly referred to as bicarbonate):

$$CO_2(g) + H_2O(l) = HCO_3^-(aq) + H^+(aq) \qquad (8.17)$$

This is an important reaction because CO_2 is released by our cells as a by-product of respiration. CO_2 is a gas at body temperature with only a limited solubility in water, and therefore it cannot be easily transported in the bloodstream. However, converting CO_2 into hydrogen carbonate, which is very soluble in water, effectively increases greatly the solubility of carbon dioxide in water. But, if our cells respire at a rapid rate, the above reaction must occur very quickly indeed if the rate of respiration is not to be limited by excessive gaseous CO_2 build-up in the blood. CA can also catalyse the reverse reaction, producing CO_2 for release at the lungs.

At pH 7, which is roughly the pH of blood serum, the uncatalysed forward reaction in Equation 8.17 occurs at a rate of about 0.1 mol^{-1} dm^3 s^{-1}: this is far too slow a conversion rate to convert all the CO_2 generated by respiration into hydrogen carbonate.

■   At pH 9, the forward reaction in Equation 8.17 is much greater at 10^4 mol^{-1} dm^3 s^{-1}. What does this suggest about the reaction mechanism?

☐   At pH 9, there is a much higher concentration of hydroxide ions in solution than at pH 7 (roughly 100 times higher). The increase in rate with OH^- concentration suggests that OH^- is involved in the rate-determining step.

In Reaction 8.17, the first step is the nucleophilic attack of either water or hydroxide on the CO_2:

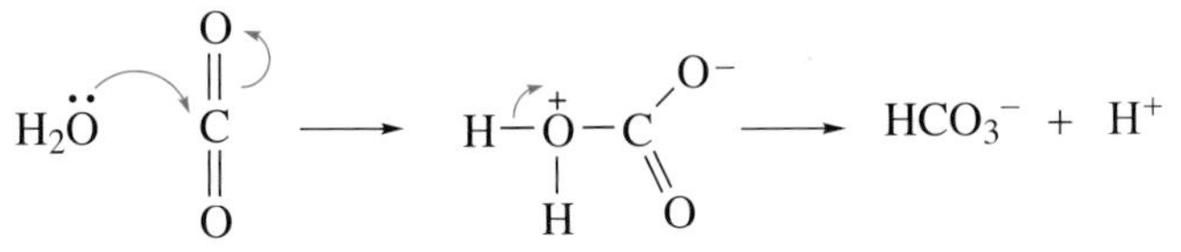

$$(8.18)$$

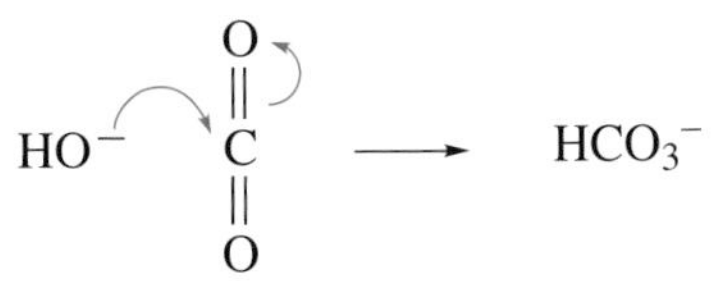

$$(8.19)$$

Formation of hydrogen carbonate is much quicker if the nucleophile is hydroxide ion (Reaction 8.19) than if it is water (Reaction 8.18). Hence, at pH 9, the reaction is much quicker than at pH 7.

As the pH of blood cannot be changed from around pH 7, Reaction 8.17 needs to be catalysed to be used as part of a CO_2 transport system in the body. Indeed, CA catalyses the reaction, such that the forward reaction now proceeds at 10^8 mol^{-1} dm^3 s^{-1} (near the diffusion-controlled limit in solution of about 10^9 mol^{-1} dm^3 s^{-1}), giving an enormous 10^9 rate enhancement over the non-catalysed reaction.

What are the structural and chemical features of CA that make it such an efficient catalyst? CA is a small/medium protein with a relative molecular mass of about 30 000. It has a roughly spherical shape, with a large cleft in the side of the sphere. The cleft contains the active site and is about 1500 pm deep (Figure 8.28a). The whole cleft is lined with hydrophobic residues. At the bottom of the cleft lies a zinc(II) ion coordinated by three histidyl side chains (Figure 8.28b). The zinc also has a water molecule (or hydroxide; see later) coordinated to it. This water molecule supports a column of hydrogen-bonded water molecules that extends out to the surface of the enzyme. Adjacent to the zinc in the active site is a group of hydrophobic amino acid residues, which forms a hydrophobic pocket.

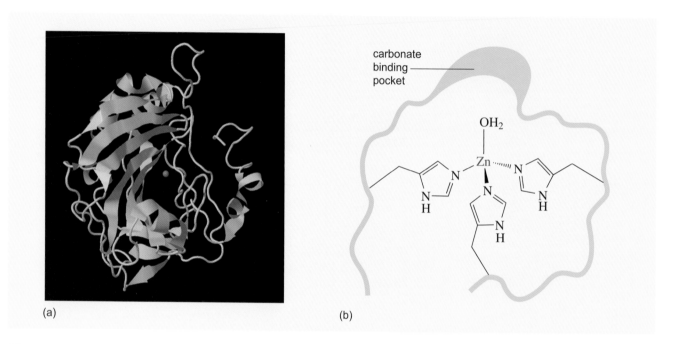

(a)   (b)

Figure 8.28   (a) Schematic diagram of the higher-order structure of carbonic anhydrase. (b) Close-up of zinc(II) active site in carbonic anhydrase. ((a) Based on pdb file 2cab (Kannan et al., 1984).)

The key to the catalytic action of CA lies in its ability to generate a reactive hydroxide ion at pH 7. Reaction 8.20 shows that the water molecule coordinated to the zinc ion readily loses a proton to give a hydroxide ligand:

$$\text{(8.20)}$$

This occurs easily because the zinc is a Lewis acid. In effect, the Lewis acidity of the zinc is translated into Brønsted acidity of the coordinated water.

■   What is the difference between a Lewis acid and a Brønsted acid?

A Brønsted acid is a proton donor, whereas a Lewis acid is defined as an electron pair acceptor.

The water coordinated to the zinc is actually considerably more acidic than normal uncoordinated water. The pK_a of the zinc-bound water is 7.4, whereas normal water has a pK_a of 15.7 (at 25 °C).

The other reactant is CO_2. The cleft acts as a funnel drawing any carbon dioxide molecules that land on the nearby enzyme surface down towards the zinc, thus increasing the rate at which CO_2 reaches the active site and contributing to the high rate of reaction we noted earlier. The CO_2 is bound in the active site next to the hydroxide ligand by the hydrophobic pocket of amino acids. As the CO_2 and nucleophilic hydroxide are brought in close proximity with each other, they react together very rapidly to give the product hydrogen carbonate. The whole catalytic cycle of CA is shown in Figure 8.29.

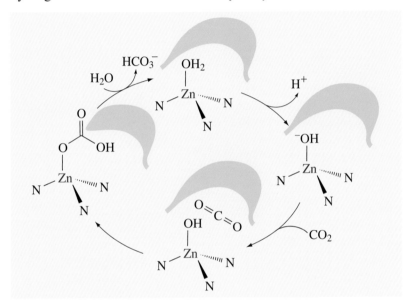

Figure 8.29 Catalytic cycle for the carbonic anhydrase-mediated conversion of CO_2 into HCO_3^-.

- Why is zinc rather than a transition metal with two stable oxidation states suitable for the active site of CA?

- The reaction mechanism involves proton transfer, not changes in oxidation state. A metal such as zinc, which is a Lewis acid, aids proton transfer by deprotonating H_2O.

8.3 Food from the air – 1: photosynthesis

Green plants obtain all their energy by **photosynthesis**, the process in which light is converted into chemical energy. Photosynthesis consists of two stages; the so called 'light' and 'dark' reactions. The light reactions are the 'photo' part of photosynthesis, where the light is converted into energy, whereas the so-called 'dark' reactions constitute the 'synthesis' part of the process. This is also called carbon fixation, using this energy to synthesise organic molecules. The light reactions require the presence of light and although the dark reactions do not require light directly, most of these take place during the day. The light and dark reactions are thus intimately linked and are summarised by the left-to-right reaction in Equation 8.21, the overall equation for photosynthesis in plants, algae and cyanobacteria. Note that (CH_2O) denotes carbohydrate in a general form.

$$CO_2 + H_2O \underset{\text{respiration}}{\overset{\text{photosynthesis}}{\rightleftharpoons}} (CH_2O) + O_2 \qquad (8.21)$$

Reaction 8.21 is of course the reverse of the consumption of carbohydrates in aerobic respiration, Equation 8.1. Reaction 8.1 is thermodynamically favourable but needs CCO and other enzymes to increase the rate of reaction. Equation 8.21 is thus thermodynamically unfavourable and it can proceed only if there is an input of energy. In photosynthesis this energy is supplied from the Sun.

In the 1940s, at the University of Illinois, Robert Emerson and co-workers provided evidence that there were two separate photosystems operating in photosynthesis (Emerson and Lewis, 1943). By measuring the yield of oxygen evolution per photon absorbed at different wavelengths, Emerson showed that the yield fell off at long wavelengths; this is known as the 'red drop' and occurs at wavelengths longer than 685 nm. However, he also found that if he irradiated the photosynthetic system at shorter wavelengths, whilst simultaneously irradiating at wavelengths greater than 685 nm, the long wavelength irradiation became effective in photosynthesis. The yield per photon absorbed of the system irradiated by both wavelengths at the same time was greater than the sum of the yields obtained by separate irradiations at the same two wavelengths. Emerson concluded, therefore, that there are two photosystems (now called **photosystem I (PSI)** and **photosystem II (PSII)**), which operate in series and have slightly different absorption spectra. At wavelengths longer than 685 nm, only PSI absorbs and PSII cannot operate, so that photosynthesis is blocked.

8.3.1   The mechanism of photosynthesis: light harvesting

Sunlight is absorbed by several pigments, of which **chlorophyll** is the most important. When a chlorophyll molecule absorbs sunlight, an electron is excited to a higher energy orbital giving chlorophyll in an excited state. The excited chlorophyll molecule then passes the energy on to another chlorophyll molecule raising the second one to the excited state. Writing chlorophyll as Chl and chlorophyll in its excited state as Chl*, we can denote this process as a chain of reactions

$$Chl_1 + h\nu = Chl_1{}^*$$

$$Chl_1{}^* + Chl_2 = Chl_2{}^* + Chl_1$$

$$Chl_2{}^* + Chl_3 = Chl_3{}^* + Chl_2$$

and so on, where the subscript numbers merely label different chlorophyll molecules.

The chain of chlorophyll molecules acts like an antenna, gathering in sunlight and channelling it towards the reaction centres where redox reactions can occur. In addition to pigments, antennae also contain structural proteins, which are different for photosystems I and II and play an important protective role: the terms light-harvesting complexes I and II (LHCI and LHCII) are often used to describe the complexes associated with PSI and PSII, respectively. Figure 8.30 shows part of the LHCII structure from a photosynthetic bacterium.

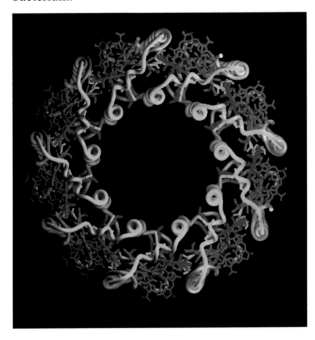

Figure 8.30 Computer-generated picture of part of the light-harvesting complex II (LHCII) structure. The supporting protein structure consists of nine pairs of α-helices (an inner ring of nine yellow helices and an outer ring of nine green helices). Sandwiched between these is a stacked ring of 18 bacteriochlorophyll a molecules (yellow and green edge-on rings). Also shown are a further ring of nine bacteriochlorophyll a molecules (blue) and nine carotenoid molecules (red).

This particular bacterium uses a chlorophyll called bacteriochlorophyll a. (*Note*: There are several different chlorophyll molecules found in living systems. We will look at the structure of the one most commonly found in plants, chlorophyll a, in Section 8.3.3. Bacteria chlorophylls are related to chlorophylls but have two extra hydrogens attached to the ring system.)

8.3.2 The overall mechanism of photosynthesis

The essential photochemical reaction is the splitting of water to form gaseous oxygen with the concurrent transfer of electrons and protons to other molecules, which therefore become reduced. To produce one molecule of oxygen, four electrons must be transferred:

$$2H_2O \longrightarrow O_2 + 4e^- + 4H^+$$

(8.22)

and four electrons are also necessary to reduce one carbon dioxide molecule to carbohydrate:

$$4e^- + 4H^+ + CO_2 \longrightarrow (CH_2O) + H_2O$$

(8.23)

Reaction 8.21 is thus the sum of Reactions 8.22 and 8.23.

In 1960, Hill and Bendall (Hill and Bendall, 1960) proposed a scheme for a two-photon system (PSI and PSII), which is still the basis of nearly all discussions of the mechanism of photosynthesis. A simple version of the scheme is shown in Figure 8.31; because of the shape of the diagram, it is usually known as the **Z scheme**. The vertical coordinate, E'_0, is a scale of electrode potentials at pH 7.

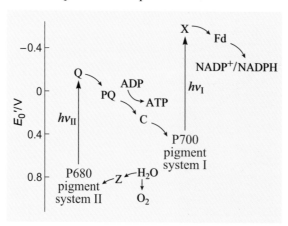

Figure 8.31 Simplified scheme of reactions involved in photosynthesis. (The abbreviations are explained in the text.)

The course of electron flow in this scheme is as follows. A quantum of light is absorbed by the pigment of PSII, exciting a group of four chlorophyll a molecules. This group is called P680 because it has an absorption peak at 680 nm. P680 provides the energy for the transfer of an electron (against the potential gradient) from a water molecule via species Z to a species Q, to give Z^+ and Q^-. Z represents a Mn_4–Ca cluster, which you will meet later in Section 8.3.4, plus a neighbouring tyrosine residue. Q represents Quinone A, a bound **plastoquinone**, **8.32**.

The electrons are transferred from P680 to Quinone A, Q, on a picosecond timescale via the intermediate electron transfer agent pheophytin a. The reduced species Q^- transfers its electron in a spontaneous dark reaction to further plastoquinone molecules (PQ) and each reduced plastoquinone transfers its electron to a complex of two cytochromes (C). There is a 'pool' of plastoquinone molecules, about 20 to each manganese ion. The (plastoquinone|reduced plastoquinone) couple appears to act as an intermediate

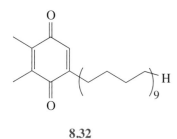

8.32

reservoir of electrons between PSII and PSI. Electrons are transferred from the cytochrome complex to PSI by the blue copper protein plastocyanin (Figure 7.17).

The reaction centre of PSI (called P700) has a lower redox potential than that of PSII, although both are thought to be groups of chlorophyll a molecules in different environments. A second photon absorbed in the PSI pigment system excites P700. Excited P700 donates an electron to another chlorophyll a molecule, which in turn donates an electron to phylloquinone (vitamin K) X. Phylloquinone donates an electron to the acceptor, which is a [4Fe–4S] cluster of the type found in ferredoxins. A second series of dark electron transfers now occurs via ferredoxin (Fd) to NADP reductase, which reduces $NADP^+$, by two electrons, to NADPH. This is the final stable product of the two-photon/electron transfer, and it acts as a reducing agent in the dark reactions that reduce carbon dioxide to carbohydrate.

■ Which of the electron transfer proteins have you met previously?

□ Cytochromes, ferredoxins and blue copper proteins were described in Chapter 7. They act as electron transfer agents in respiration (Section 8.2.1) as well as in photosynthesis.

In the following sections, we will only consider the metal-containing centres in the light reactions – chlorophyll and the manganese clusters.

8.3.3 Chlorophyll

Why is chlorophyll a good molecule for the absorption of sunlight? What properties are needed for a useful absorber of sunlight?

The molecule should have a strong absorption (high ε) over a range of wavelengths corresponding to solar radiation reaching the Earth's surface. Figure 8.32 compares the spectrum of chlorophyll a with the intensity distribution of solar radiation at the Earth's surface. Infrared radiation does not provide enough energy to overcome the unfavourable ΔG values and the intensity of ultraviolet radiation is reduced by absorption in the upper atmosphere. A strong absorber of visible radiation is therefore needed. Figure 8.32 shows that chlorophyll a absorbs strongly in two regions of the visible spectrum.

■ Why does the absorption spectrum of chlorophyll a explain why this molecule is green?

□ There are strong absorption bands in the red (around 700 nm) and blue (around 400 nm) regions of the spectrum. A molecule absorbing red and blue will appear green.

An ideal photoabsorber would absorb strongly over the entire visible region and appear black.

The general structure of chlorophyll is shown in **8.33**. The side chain R varies for the different chlorophylls. In chlorophyll a, R is **8.34**.

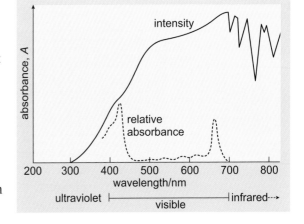

Figure 8.32 Relative absorbance of chlorophyll a and the intensity distribution of solar radiation at the Earth's surface.

8.33

CH_2

8.34

By now, you will probably recognise that the ligand, **chlorin**, surrounding the magnesium ion is related to the porphyrins. In chlorin, one of the five-membered rings in porphyrin is replaced by two fused five-membered rings, which link directly to another five-membered ring.

■ In corrin (Section 7.3) direct linking of two five-membered rings led to a break in delocalisation. Does the same happen here?

☐ No. A continuous path of alternate double and single bonds can be traced through the chlorin ring.

This delocalised system is important for chlorophyll's role in photosynthesis.

■ Some transition metal proteins, for example blue copper proteins, are coloured as a result of d–d or charge transfer transitions. Are these types of transitions likely to be responsible for the colour of chlorophyll?

☐ No. Mg^{2+} is not a transition metal ion. It has an inert gas configuration [Ne] and a transition to an excited state involving Mg^{2+} is likely to involve an energy corresponding to ultraviolet radiation because an electron would have to be promoted from the $n = 2$ shell to the $n = 3$ shell.

The transitions giving rise to the green colour of chlorophyll are those involving delocalised orbitals in the ligand.

So what is the role of the magnesium ions? Although the Mg^{2+} ion does not absorb the radiation itself, it does affect the electron distribution and orbital energies in the chlorin ring; as a consequence it leads to an increase in wavelength and extinction coefficient, ε. The variation in wavelength absorbed for different chlorophylls is due to fine-tuning of the energy levels through slight changes in the coordination of Mg. The lack of absorption by magnesium itself may also be an advantage as it means that there is no interference with the absorption by the chlorin ring.

8.3.4   The Mn_4Ca cluster

As we stated earlier, the role of chlorophyll is to absorb energy from sunlight and channel it to the active sites so that it can be used to drive the oxidation of water. We now go on to study the catalytic centre where the actual oxidation of water occurs.

The catalytic centre in photosystem II for the oxidation of water is a cluster containing four manganese ions and a calcium ion.

■ It was noted in Section 8.3.2 that the oxidation of water is a four-electron process. What property of manganese would suggest it is suitable to be involved in such a reaction?

☐ Manganese has a wide range of stable oxidation states from +2 to +7.

At the time of writing (2009), the structure of this cluster is still a matter of debate. We shall use this cluster as an illustration of the difficulty of determining unambiguous structures and how a picture of the structure can be built up from the results of several techniques.

The structure of the Mn cluster has been studied by a variety of techniques. Analysis of the Mn content found that there were four Mn ions per active site. Mn–EXAFS examination of the cluster indicated O or N atoms at around 180 pm, and Mn at 270 pm and 330 pm. The Mn–Mn distances of 270 pm and 330 pm are typical of oxygen-bridged Mn distances, corresponding to $\mu-\eta^1$, η^1 coordination of two and one oxygen atoms, respectively. The results here are harder to interpret than results on single-centre complexes as the EXAFS peaks refer to average distances from the four Mn ions.

■ The Mn–EXAFS results reported O or N atoms. Why is it difficult to distinguish N and O with this technique?

☐ The scattering of X-rays is determined by the electron density around the atom. For atoms close together in the Periodic Table, such as N and O, the densities will be similar. This is also a problem to some extent for X-ray diffraction.

In 1984/5, the X-ray structure of a related system, the purple bacterial reaction centre was determined (Deisenhofer et al., 1984). Purple bacteria have a single photosystem, which behaves similarly to PSII. After sequencing of PSII, it was found that parts of the purple bacteria unit sequence were similar to part of PSII. Thus, the known X-ray structure of the bacterial system could be used as a starting point to construct models of PSII.

The X-ray structure of the Mn cluster in PSII has since been determined but only at medium resolution – 380, 350 and 300 pm. This is insufficient to fix the atom positions exactly as only atoms 300 pm or more apart could be distinguished. As the EXAFS results show, there are atoms 180 pm away from the Mn and their positions cannot be determined accurately. However other evidence combined with the X-ray structure enabled a picture of the cluster to be deduced (Loll et al., 2005). This is shown in Figure 8.33a.

This rather unusual looking arrangement is essentially a distorted tetrahedron formed from 3Mn + Ca, with the fourth Mn linked to one of the other Mn ions. This was unexpected as 350 pm resolution X-ray data (Ferreira et al., 2004) had led to a proposal of a cubane structure similar to Fe–S clusters. However it has also been pointed out that the X-rays used to obtain the diffraction pattern can damage the protein (Yano et al., 2005). For the Mn_4–Ca cluster, X-ray spectroscopy showed that the manganese in one of its intermediate oxidation states, the S_1 state (see the next section) was reduced from (2Mn(III) + 2Mn(IV)) to 4Mn(II) by an order of magnitude lower dose of X-rays than was used for the crystal structure determination. Data from a more recent EXAFS experiment (Kern et al., 2007) could not be made to fit either structure. The best fit was to a model with three coupled Mn–Mn dimers with Mn–Mn distances of

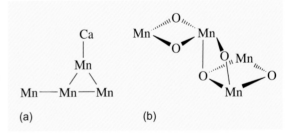

(a)   (b)

Figure 8.33 (a) Schematic view of the Mn_4–Ca cluster as determined by 300 pm resolution X-ray diffraction showing just the metal ions. (b) Structure of the Mn_4 part of the cluster as suggested by EXAFS.

270–280 pm and an Mn–Mn distance of 300 pm between Mn ions in two separate dimers. Four possible structures including this model were identified but could not be distinguished. One of these structures is shown in Figure 8.33b.

8.3.5 The photocatalytic process

We noted that the wide range of oxidation states of Mn was an advantage for its role in water oxidation but which oxidation states are actually used in PSII?

The cluster is oxidised in a series of oxidation steps losing one electron at a time in response to excitation of P680. Five states have been observed spectroscopically and are labelled S_0, S_1, S_2, S_3 and S_4. The cycle in which these states are produced and water oxidised is shown schematically in Figure 8.34.

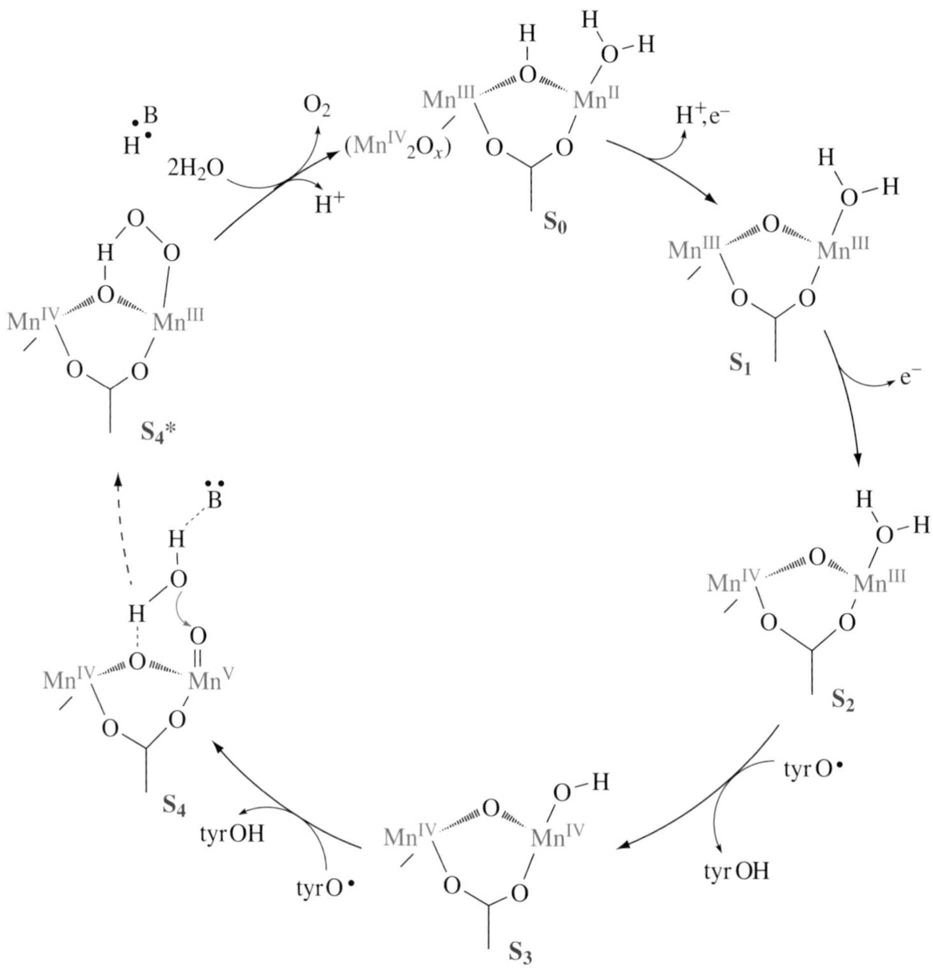

Figure 8.34 Photocatalytic cycle for the oxidation of water by the Mn cluster. tyr represents a tyrosine residue and B a Lewis base. The $(Mn^{IV}_2O_x)$ group is unaltered during the cycle and so is not shown other than in the S_0 state.

The oxidation states of manganese in each of the five states are thought (2007) to be those shown in Table 8.1.

Table 8.1 The oxidation states of Mn in the spectroscopically observed states.

State	Oxidation state(s)
S_0	$2 \times$ Mn(IV), $1\times$ Mn(III), $1 \times$ Mn(II)
S_1	$2 \times$ Mn(IV), $2 \times$ Mn(III)
S_2	$3 \times$ Mn(IV), $1 \times$ Mn(III)
S_3	$4 \times$ Mn(IV)
S_4	$3 \times$ Mn(IV), $1 \times$ Mn(V)?

XANES studies (Britt, 1996) indicate that the most likely oxidation states for Mn in the state S_1 are two Mn^{4+} ions and two Mn^{3+} ions. UV–visible spectroscopy has indicated changes in oxidation state of Mn^{2+} to Mn^{3+} and Mn^{3+} to Mn^{4+}. The electronic transitions producing the spectral lines are thought to be charge transfer transitions and occur at wavenumbers typical for such transitions for the ions proposed. The oxidation states of S_4 are the most uncertain. Proposals include an S_4 state with one Mn oxidised to Mn^{5+} and forming an Mn=O group and S_4 as S_3 but with an electron removed from the tyrosine residue, to form a radical.

- Where have you previously met a possible intermediate state with an electron removed from groups surrounding the metal ion to give a radical?

- The reactions catalysed by P450s (Section 8.2.4) and peroxidases (Section 8.2.5) could both have an intermediate with an Fe(IV)=O bond and an electron removed from the surrounding porphyrin or an amino acid residue (suggested by spectroscopic studies).

Studies of model Mn complexes containing two or four manganese ions linked by bridging oxygen showed that these complexes reduced water to O_2 in the presence of certain oxidising agents (Kurz et al., 2007). Oxygen evolution was not detected for good oxidising agents with high electrode potentials but containing no oxygen, for example Ce^{4+}; however, a high yield of oxygen was observed for oxidants that could transfer oxygen to the metal. The researchers concluded that an Mn=O group was likely to be an intermediate in the photocatalytic cycle.

The simple loss of an electron at each stage would lead to a build-up of charge on the cluster and this would greatly affect how the cluster would interact with its surroundings. Most models therefore propose that the electron loss is accompanied by the loss of a proton.

Measurements of rates of exchange of free water with water bound to manganese indicate that the first water molecule needed for the reaction is bound to S_0 but that the second does not bind until S_3.

The roles of Ca^{2+} and Cl^- (which is also found in PSII) are uncertain.

8.4 Food from the air – 2: nitrogen fixation

Nitrogen is an essential element needed to manufacture many biological molecules such as proteins, RNA, DNA and ATP.

Nitrogen is not, of course, in short supply; it is the fourth most abundant element in the biosphere, and constitutes nearly 80% of the atmosphere. Unfortunately in the atmosphere it is present as dinitrogen, N_2, and dinitrogen is the most inert of diatomic molecules – or should we say fortunately, as we should not otherwise exist. Our atmospheric oxygen would have combined with some of the nitrogen, if the activation energy were not so high, to form oceans of nitric acid, as nitrate formation is thermodynamically favourable. (It is just as well that nature has not found an efficient catalyst for this reaction.) The only living organisms that can use dinitrogen directly are the small proportion of bacteria that make the **nitrogenase** enzyme. Animals and plants cannot fix nitrogen; plants rely on the products from nitrogen fixation to provide their nitrogen requirements. For example in Chapter 3 (Table 3.3), you saw that plants absorb nitrogen in the form of nitrate, NO_3^-, or ammonium, NH_4^+, ions.

Figure 8.35 illustrates the nitrogen cycle in the biosphere. Nitrogen-fixing bacteria, produce (mostly) ammonia from N_2 in the air. This is taken up by plants. Animals (fish, insects, etc.) eat plants and other animals. The animals (and bacteria) return nitrogen compounds to the soil and water as waste products. The nitrogen compounds, in turn, are consumed by other bacteria and plants.

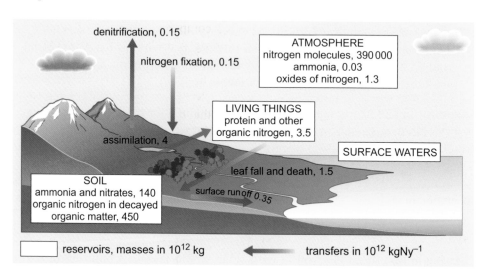

Figure 8.35 The nitrogen cycle.

Thus, ammonia left in the soil by the decay or 'ammonification' of organic matter (particularly amino acids) may be used as an energy source by so-called nitrifying bacteria, and oxidised to nitrite or nitrate. The nitrate ion is more mobile in soils than ammonia or ammonium ions, which are trapped by clay particles; so nitrate reaches the roots more readily, and may then be reduced back to ammonia by the plant. Much soil nitrate, however, can be reduced by (so-called) denitrifying bacteria, which use nitrite or nitrate as a source of oxygen, producing N_2O or N_2 as byproducts; and much is washed away.

Nitrogen-fixing bacteria may be free living in air, soil or water, or symbiotic with plants. The bacteria of greatest importance to agriculture are the rhizobia, which colonise legumes, forming nodules on their roots. Legumes include weed-like plants such as clover, crops such as alfalfa (lucerne) used for animal fodder, beans and peas, flowering plants such as lupins, and large trees such as acacia. One example, cyanobacteria (formerly called blue–green algae) are also important because of their ability to photosynthesise. Some lichens, which are symbioses (combinations) of fungi and algae, include nitrogen-fixing cyanobacteria, and some even use cyanobacteria for photosynthesis instead of green algae.

Because of the problems of reducing dinitrogen, nitrogenase is a complicated enzyme, working slowly, and requiring a large energy input. Consequently it has not achieved the high specificity of other enzymes. For example, ethyne, HC≡CH, is reduced preferentially by the enzyme, and NO and CO inhibit reduction of N_2. Dioxygen destroys nitrogenase irreversibly, and although leguminous plants maintain a low concentration of free oxygen in their roots, nitrogen-fixing bacteria must protect the enzyme whilst obtaining the oxygen they need for other processes.

8.4.1 Coordination chemistry of the dinitrogen ligand

Surprisingly, the unreactive dinitrogen molecule can form transition-metal complexes under mild conditions. Table 8.2 compares properties of some dinitrogen complexes with those of the dinitrogen molecule, the transient cation N_2^+ (observed in comets and in the upper atmosphere), and other simple compounds with NN bonds. It shows how the NN bond length, $r(NN)$, increases and the NN stretching frequency, $v(NN)$, decreases as the bond order decreases from 3 in N_2 to 2.5 in N_2^+. From this comparison we can deduce that the NN bond order in the ruthenium complexes is about 2.5.

Table 8.2 Properties of dinitrogen compounds, ions and complexes.

Compound	$v(NN)/cm^{-1}$	$r(NN)/pm$	$r(RuN)/pm$
N_2	2331	109.8	–
N_2^+	2175	111.8	–
EtN=NEt	1576	123	–
$H_2N–NH_2$	1098	145	–
$[Ru(NH_3)_5(N_2)]Cl_2$	2130	≈112	≈210
$[\{Ru(NH_3)_2(\mu^2\text{-}N_2)\}](BF_4)_4$	2100	112	193

It is a general observation that terminal dinitrogen complexes have NN bond lengths and stretching frequencies that are intermediate between the values for triple and double NN bonds (depending on the metal and the coligands). The bond in the ligand is weaker than in free N_2 because of donation of metal electrons into the ligand π^* orbital. The metal–ligand bonding is qualitatively the same as metal–CO bonding. However the bonding is much weaker for N_2.

Why should this be, given the similarities of the N_2 and CO molecules? The energies of the valence orbitals are similar for N_2 and CO, as might be expected, because nitrogen lies between carbon and oxygen in the Periodic Table. But the energy of the σ-donor orbital of N_2 is relatively low, and it is not as good a match for the d orbitals of most metals as the corresponding orbital of CO. Dinitrogen is thus a relatively poor σ-donor and a moderate π-acceptor.

Table 8.3 shows a range of terminal dinitrogen complexes for some different metals with coordination numbers 4, 5 and 6. As appropriate for π-acceptor ligands, the metal oxidation numbers are relatively low, from -1 to $+2$ (with higher numbers for metals to the right of the transition series). All have lower NN stretching frequencies than in the N_2 molecule, depending on the metal and other ligands in the complex (coligands), and hence longer NN bonds (where the structures have been measured). The nitrogen NMR shifts are not very different from the N_2 value, so that the change in electronic structure may not be very great. The shielding is usually higher for the ligating than the non-ligating nitrogen, but not always.

Most dinitrogen complexes are low spin (as N_2 is a strong-field ligand) and obey the 18-electron rule. (Table 8.3 contains a rather high proportion (one in five) of complexes that do not, so as to illustrate the potential variety of the complexes.) Most have octahedral coordination, but other geometries – for example, square planar and tetrahedral – are represented also.

Table 8.3 Properties of some terminal dinitrogen complexes.

Complex	$\nu(NN)/\text{cm}^{-1}$	Colour	$r(NN)/\text{pm}$	$\delta(^{15}N_\alpha)$	$\delta(^{15}N_\beta)$
$[Et_4N][V(CO)_5(N_2)]$	1843	red–violet	—	—	—
$[Cr(\eta^6\text{-}C_6H_6)(CO)_2(N_2)]$	2145	red	—	—	—
$[Mo(N_2)(PMe_3)_5]$	1950	orange	112	-35	-41
trans-$[Mo(N_2)_2(dppe)_2]$	1979	yellow–orange	112	-43.1	-42.8
trans-$[Mo(N_2)_2(dppe)_2]Cl$	2043	red	—	—	—
trans-$[W(N_2)_2(dppe)_2]$	1948	orange	—	-60	-49
$[Mn(CO)_2(\eta^5\text{-}C_5H_5)(N_2)]$	2169	red–brown	—	—	—
trans-$[ReCl(N_2)(dppe)_2]$	1980	yellow	—	-92	-65
trans-$[ReCl(N_2)(dppe)_2]Cl$	2060	purple	—	—	—
$[FeH_2(N_2)(PPh_3)_3]$	2008	yellow	—	—	—
$[Ru(NH_3)_5(N_2)_3]Cl_2$	2130	yellowish	112	-81	-44
$[Os(N_2)oep]$ *	2030	dark violet	—	—	—
$[CoH(N_2)(PPh_3)_3]$	2090	red–orange	116	—	—
$[Ni(N_2)(Pet_3)_3]$	2070	purple	—	—	—
cf. N_2	2331		109.8		-75

* oep = octaethylporphyrin (cf. haem, but with all the R groups ethyl).

8.4.2 Constitution of nitrogenase

Four distinct nitrogenase systems are known. Three of these are closely related and are binary enzymes consisting of two proteins, referred to as the nitrogenase complex. Protein 1 is a molybdenum-, vanadium- or iron-based nitrogenase. Protein 2 is an iron-based reductase. All nitrogen-fixing bacteria using one of these three systems contain the molybdenum-based nitrogenase, but some bacteria contain the vanadium-based system or the iron-based system as well or even all three. We shall concentrate on the molybdenum-based nitrogenase, which has been the most extensively studied. Protein 1 in this case is known as the **MoFe protein**. Protein 2 is known as the **Fe protein**.

Why molybdenum?

The bioactivity of molybdenum is remarkable, as it is a very rare element in the Earth's crust – 50 000 times less abundant than iron. The sequence of natural abundance for transition metals in the Earth's crust begins:

$$Fe \sim Ti > Mn > Zr > V > Cr > Ni > Zn > Cu > Co > U > Mo \sim W$$

Yet molybdenum is the only metal in the second or third transition series that is known to be essential to life.

Clues to the bioactivity of molybdenum are related to its fairly central position in the transition series and its ability to form stable compounds in a wide range of oxidation states.

■   Given that Mo is the same Group of the Periodic Table as Cr, what would you expect its highest oxidation state to be?

□   +6 corresponding to the removal of the $5s^2$ and $4d^4$ electrons.

In fact molybdenum displays a range of oxidation states from −2 to +6, and coordination numbers from 4 to 8 or 9 (with small ligands). Some examples are shown in structures **8.35**–**8.39** (where X is a halogen).

It is significant also that molybdenum(VI) oxide (MoO_3) and molybdenum(IV) sulfide (MoS_2), often with the addition of other metals, are important industrial hydrogenation catalysts. Sulfur and nitrogen are removed from crude petroleum in this way: H_2 is adsorbed onto the catalyst, and dissociates to form thiolate (–SH) ligands, which promote C–S or C– N bond cleavage (hydrogenolysis) in organic compounds. Molybdenum catalysts are important also for the hydrogenation of coal to oil, and for dehydrogenations.

It has been suggested that molybdenum may have acquired its role in enzymes in the sea, where life probably began, and in which molybdenum is the most abundant transition metal (Table 3.1).

Life is thought to have begun in a reducing or neutral atmosphere, with sulfur (and sulfur bacteria) playing a greater role than now; so the molybdenum may have been associated with sulfur, as in the ore molybdenite, MoS_2, and, as you will see, in nitrogenase.

8.35

8.36

8.37

8.38

8.39

The primeval reducing atmosphere may, however, have been rich in ammonia, in which case nitrogenase would not have been needed until oxygen came to dominate. Nitrogenase might then have developed from a more primitive enzyme with a detoxifying function, reducing cyanide (as nitrogenase does).

In our oxygen containing atmosphere today, molybdenum is normally available at pH 7 as molybdate, MoO_4^{2-}, which is water-soluble (as indeed is thiomolybdate, MoS_4^{2-}). It is unusual for plants to take up metals in anionic form: contrast the availability of the cations Fe^{2+}, Fe^{3+}, Co^{2+}, Ni^{2+} and Cu^{2+}. Molybdate, however, can be absorbed with sulfate or phosphate.

■ What is the oxidation state of Mo in MoO_4^{2-}?

□ +6.

Molybdenum is found not only in nitrogenase but also in a range of other enzymes; it is an essential trace element in our diet. In humans there are several Mo-containing enzymes in the liver. Most of these enzymes catalyse reactions in which substrates are oxidised. For example, the aldehydes produced by LADH (see Section 7.2) are oxidised by the Mo-containing enzyme aldehyde oxidase to carboxylic acids.

8.4.3 The molybdo (or MoFe) protein

The MoFe protein is a tetramer of two types of subunit, $\alpha_2\beta_2$, with a molecular weight of about 230 000. It contains two types of metal–sulfur clusters, the **P cluster** and the **FeMo-cofactor**. Each $\alpha\beta$-unit contains one of each type. Although MoFe protein is often treated as a dimer of dimers, an active single $\alpha\beta$ unit has not been isolated.

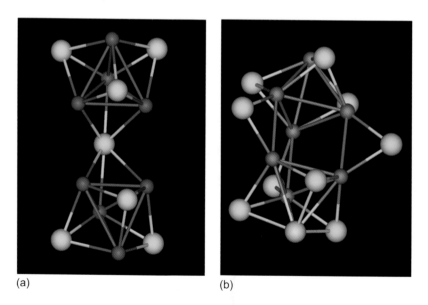

(a)   (b)

Figure 8.36 (a) P cluster in the MoFe protein of nitrogenase. (b) Structure of the FeMo-cofactor. (Yellow represents sulfur, brown is iron and blue is molybdenum.) (Based on pdb files (a) 1min clf and (b) 1min clm (Peters et al., 1997).)

The P cluster is an [8Fe–7S] cluster, which is composed of two subclusters: a [4Fe–4S] cluster linked by one of its S atoms to a [4Fe–3S] cluster, Figure 8.36a. The shared S ion is coordinated by six Fe ions. The clusters are located at the interface of the α- and β-subunits with cysteinyl residues from the α-subunit coordinated to the [4Fe–4S] cluster and cysteinyl residues from the β-subunit coordinated to the [4Fe–3S] cluster. In addition one cysteinyl residue from each subunit forms a sulfide bridge between the two subclusters.

■ Where have you met Fe–S clusters before?

□ Such clusters occur in ferredoxins (Section 7.4.3).

The FeMo-cofactor cluster also consists of two subclusters. One of these is a [4Fe–3S] cluster, the other is also a cubane Fe–S cluster with one S missing but in this case one Fe is replaced by Mo [Mo–3Fe–3S], Figure 8.36b. These two subclusters are linked by bridging non-protein-based sulfides. In the centre is a light atom. The identity of this atom is still a matter of controversy. It had been generally thought this was an N atom but spectroscopic studies combined with calculations suggest that the atom is unlikely to be N or C and O has been considered as a possibility. The FeMo-cofactor is located 100 pm below the protein's surface and there is evidence to suggest this is the N_2-binding site where reduction of N_2 to NH_3 occurs.

8.4.4 The Fe protein

As we stated in 8.4.2, the second protein of nitrogenase, the iron containing reductase, is known as the Fe protein. This is a smaller protein, molecular mass 64 000, consisting of two subunits, Figure 8.37.

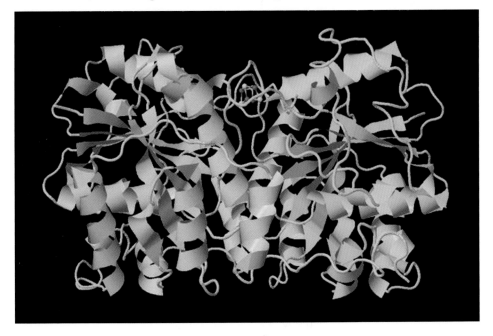

Figure 8.37 Structure of the Fe protein. (Based on pdb file 1g5p (Strop et al., 2001).)

The two subunits are attached via cysteinyl residues to a bridging [4Fe–4S] cluster. The cluster is at one end of the interface between the two subunits where it is exposed to solvent.

An unexpected but crucial aspect of the structure is that each subunit has a site where nucleotides can bind. In the oxidised form, MgADP (that is adenosine diphosphate plus a magnesium ion) is bound across the two subunits.

8.4.5 Mechanism of nitrogenase

The complete mechanism of nitrogen fixation is still not clear. However, the overall process is shown in Figure 8.38.

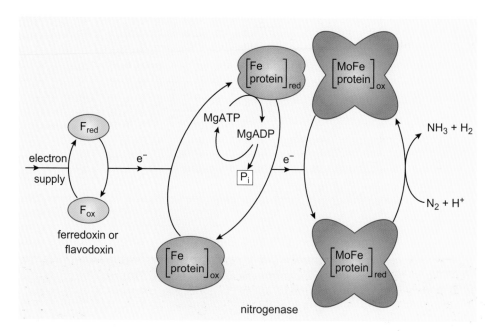

Figure 8.38 Scheme of the general pathway through nitrogenase.

Electrons are delivered to the Fe protein by a ferredoxin or flavodoxin. The [4Fe–4S] cluster is reduced from $[4Fe–4S]^{2+}$ to $[4Fe–4S]^{+}$ with the energy required provided by the hydrolysis of ATP (reacting to form ADP).

■ What is the average oxidation state of iron in the oxidised and reduced clusters?

☐ Counting S as −2, in the oxidised state the average oxidation state of Fe is 10/4 = 2.5 and in the reduced state it is 9/4 = 2.25.

One way of looking at this is that the oxidised cluster contains two Fe^{2+} ions and two Fe^{3+} ions and the reduced cluster contains three Fe^{2+} ions and one Fe^{3+} ion.

Although the exact mechanism is unclear, the current view is that reduced Fe protein delivers an electron to the P cluster which is then delivered to the FeMo-cofactor and thence to the substrate. The X-ray structure of the nitrogenase system shows that the P cluster is midway between the Fe–S cluster of the Fe protein and the FeMo-cofactor thus supporting this electron transfer route. Spectroscopic observations however failed to detect any change in the P cluster during the reduction of nitrogen, but this may be due to the speed at which the electron transfer occurs.

As we implied in Section 8.4.1, it has always been assumed that N_2 attaches to a metal ion. Evidence for this has been difficult to obtain mainly because N_2 (and indeed other substrates and inhibitors) only binds if the Fe protein, the MoFe protein, MgATP and a reductant are all present. When all these are present, however, the reaction occurs rapidly so it is difficult to follow. One piece of evidence in favour of N_2 binding to the FeMo-cofactor comes from a study of mutant bacterial strains, which are unable to biosynthesize FeMo-cofactor. The mutant strains cannot fix nitrogen, but this ability is restored when FeMo-cofactor is added to crude extracts from the mutant strains. A second line of evidence comes from studies (George et al., 1997) of nitrogenase in the presence of carbon monoxide. CO inhibits the reduction of N_2 but does allow the reduction of H^+ to H_2. At high CO concentrations, two CO stretching vibrations at frequencies characteristic of terminal CO are observed in the infrared spectrum. At lower CO concentrations, one vibration corresponding to bridging CO is observed. Other spectroscopic techniques (Lee et al., 1997) have indicated that CO is attached to Fe ions in the FeMo-cofactor. It is thus suggested that at high concentrations, CO attaches to two separate Fe atoms in the FeMo-cofactor and at low concentrations, it forms a bridge between them. As CO blocks the action of nitrogenase, it is likely that the molecule is occupying the site that should be occupied by N_2 and hence N_2 attaches to Fe.

In the current proposed scheme, Figure 8.39, the binding and reduction of N_2 requires the accumulation of eight electrons by the MoFe protein in single-electron transfer steps. At each stage proton transfer accompanies electron transfer.

$MoFe_0$ is the resting state of the MoFe protein. The model involves the accumulation of three electrons and three protons in three single-electron transfer steps before N_2 is bound in step **4**. Step **4** is a two-stage process involving transfer of an electron and a proton to the MoFe protein and binding of N_2 coupled to loss of H_2. These two stages can occur in either order as indicated by **4a** and **4b**. N_2 is bound as $M–NNH_2$. Successive additions of proton and electron then lead to $M=NNH_3$, $M=NH$, MNH_2 and finally loss of the second nitrogen atom as NH_3.

- ■ Assuming that the reducing agent is a proton, write an equation for the reduction of N_2 to NH_3 and hence deduce the number of electrons required for this process.

- □ $N_2 + 6H^+ + 6e^- \rightarrow 2NH_3$

Thus six electrons are needed.

- ■ Why does the overall process require 8 electrons?

- □ In step **4** H_2 is produced, the overall reaction being $2H^+ + 2e^- \rightarrow H_2$.

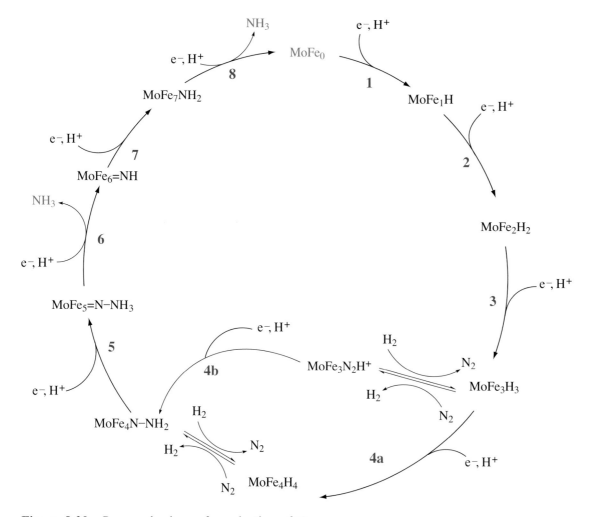

Figure 8.39 Proposed scheme for reduction of N_2.

■ At what stage is the first nitrogen atom lost as NH_3?

□ In the conversion of M=NNH$_3$ into M=NH, step **6**, that is on the transfer of the sixth proton and electron to the MoFe protein.

It is not known where the protons and electrons are stored in the MoFe protein.

8.5 Endnote – metalloproteins as metal complexes

In this chapter and the previous one, we have covered several processes all of which involve proteins with metal ions at the active site. Rather than thinking of the proteins as large biomolecules which happen to contain a metal ion, we can think of them as metal complexes with unusually large ligands. We finish this chapter by looking at the roles these ligands play.

In general, changing the ligands around a metal ion will change the electrode potential for the oxidation or reduction of that ion, making it more or less favourable for the reaction to occur and in some cases changing the relative stability of different oxidation states. In metalloenzymes this has evolved as a fine art. By subtle alterations of the groups attached to ligating ring systems or changes in ligating amino acid residues, the electrode potential can be made to match that of the process being catalysed. This is especially evident in the electron transfer proteins. Highly distorted geometries are often also a feature of biosystems and this has a direct bearing on the reactivities of the metals, both in terms of redox potential and in terms of substrate binding specificity. This leads, in a variety of systems, to the stabilisation of unusual oxidation states which would not be readily achieved in simpler reactions in the laboratory. For example, Fe(IV) in CCO, catalases and peroxidases and Co(I) in coenzyme B_{12} (Section 7.3).

The ligands are often rigid so that there is little change in geometry on reaction. This saves on reorganisation energy. One common ligand arrangement is an almost flat ring system as in the haem and related groups. Some systems indeed manifest an environment around the metal that is an ingenious compromise of the preferred geometry of two oxidation states, for example blue copper proteins (Section 7.4). Another advantage of rigidity is that stable coordinatively unsaturated complexes can be formed. They provide a vacant site where reacting molecules, for example O_2, can be attached.

The ligand can also protect the metal centre from molecules that might interfere with the reaction taking place at the centre. The amino acids in nitrogenase, for example, bury the metal centre where O_2 cannot reach it. In haemoglobin and myoglobin, the ligand provides an environment that sterically prevents dimerisation, ensuring reversible binding of oxygen.

In addition, the ligand environment beyond the coordination sphere of the metal centre can provide a pathway for electrons or protons from one site to another. For example, the electrons are transported from the external site (to which electron transfer proteins deliver them) to an internal, protected site where reaction occurs, as in the proposed pathway for the cytochrome c/cytochrome c peroxidase (Section 7.4.1, Figure 7.12).

Thus, although most of the reactions catalysed by the enzymes we have discussed occur at the metal centre, the surrounding protein as a whole is vitally important in modifying the course or rate of the reaction to fulfil its biological role.

This chapter has considered how naturally occurring ligands in proteins are finely tuned for their purpose. In Chapter 9 we shall see how researchers have designed ligands to modify the properties of metal complexes for medical applications.

Don't forget that there are questions on the companion website which you can use to test your understanding of the material covered in this chapter.

References

Banerjee, R. (2003) 'Radical carbon skeleton rearrangements. Catalysis by coenzyme B_{12}-dependent mutases', *Chemical Reviews*, vol. 103, pp. 2083–94.

Bertini, I., Bryant, D.A., Ciurli, S., Dikiy, A., Fernández, C.O., Luchinat, C., Safarov, N., Vila, A.J. and Zhao, J. (2001) 'Backbone dynamics of plastocyanin in both oxidation states. Solution structure of the reduced form and comparison with the oxidized state', *Journal of Biological Chemistry*, vol. 276, no. 50, pp. 47217–26.

Bialek, W., Krzywda, S., Jaskolski, M. and Szczepaniak, A. (2009) 'Atomic-resolution structure of reduced cyanobacterial cytochrome c6 with an unusual sequence insertion', *Federation of European Biochemical Societies Journal*, vol. 276, pp. 4426–36.

Braun, W., Vasak, M., Robbins, A.H., Stout, C.D., Wagner, G., Kagi, J.H. and Wuthrich, K. (1992) 'Comparison of the NMR solution structure and the X-ray crystal structure of rat metallothionein-2', *Proceedings of the National Academy of Sciences of the United States of America*, vol. 89, no. 21, pp. 10124–8.

Britt, R.D. (1996) 'Oxygen Evolution', pp. 137–64, in *Oxygenic Photosynthesis: The light reactions*, Ort, D.R. and Yocum, C.F. (eds), Dordrecht, The Netherlands, Kluwer Academic Publishers.

Buchanan, S.K., Smith, B.S., Venkatramani, L., Xia, D., Esser, L., Palnitkar, M., Chakraborty, R., van der Helm, D. and Deisenhofer, J. (1999) 'Crystal structure of the outer membrane active transporter FepA from *Escherichia coli*', *Nature Structural Biology*, vol. 6, pp. 56–63.

Burke, J.M., Kincaid, J.R., Peters, S., Gagne, R.R., Collman, J.P. and Spiro, T.G. (1978) 'Structure-sensitive resonance Raman bands of tetraphenyl and "picket fence" porphyrin-iron complexes, including an oxyhemoglobin analogue', *Journal of the American Chemical Society*, vol. 100, pp. 6083–8.

Debye, P. and Huckel, E. (1923) 'The theory of the electrolyte II – The border law for electrical conductivity', *Physikalische Zeitschrift*, vol. 24, pp. 304–25.

Deisenhofer, J., Epp, O., Miki, K., Huber, R. and Michel, H. (1984) 'X-ray structure analysis of a membrane protein complex. Electron density map at 3Å resolution and a model of the chromophores of the photosynthetic reaction center from *Rhodopseudomonas viridis*', *Journal of Molecular Biology*, vol. 180, pp. 385–98.

Doyle, D.A., Morais Cabral, J., Pfuetzner, R.A., Kuo, A., Gulbis, J.M., Cohen, S.L., Chait, B.T. and MacKinnon, R. (1998) 'The structure of the potassium channel: molecular basis of K^+ conduction and selectivity', *Science*, vol. 280, pp. 69–77.

Duer, M. (2008) personal communication.

Emerson, R. and Lewis, C.M. (1943) 'The dependence of the quantum yield of chlorella photosynthesis on wave length of light', *American Journal of Botany*, vol. 30, pp. 165–78.

Ferreira, K.N., Iveson, T.M., Maghlaoui, K., Barber, J. and Iwata, S. (2004) 'Architecture of the photo-synthetic oxygen-evolving center', *Science*, vol. 303, pp. 1831–8.

Flexman, A.M., Del Vicario, G. and Schwarz, S.K.W. (2007) 'Dark green blood in the operating theatre', *The Lancet*, vol. 369, p. 1972.

Frank, F.C. (1949) 'The Influence of dislocations on crystal growth', *Faraday Society Discussions*, vol. 5, p. 48.

George, S.J., Ashby, G.A., Wharton, C.W. and Thorneley, R.N.F. (1997) 'Time-resolved binding of carbon monoxide to nitrogenase monitored by stopped-flow infrared spectroscopy', *Journal of the American Chemical Society*, vol. 119, pp. 6450–1.

Hersleth, H.P., Uchida, T., Røhr, A.K., Teschner, T., Schünemann, V., Kitagawa, T., Trautwein, A.X., Görbitz, C.H. and Andersson, K.K. (2007) 'Crystallographic and spectroscopic studies of peroxide-derived myoglobin compound II and occurrence of protonated FeIV O', *Journal of Biological Chemistry*, vol. 282, no. 32, pp. 23372–86.

Hill, R. and Bendall, F. (1960) 'Function of the 2 cytochrome components in chloroplasts –working hypothesis', *Nature*, vol. 186, pp. 136–7.

Iwata, S., Ostermeier, C., Ludwig, B. and Michel, H. (1995) 'The structure at 2.8-Angstom resolution of cytochrome c oxidase from paracoccus denitrificans', *Nature*, vol. 376, pp. 660–9.

Kannan, K.K., Ramanadham, M. and Jones, T.A. (1984) 'Structure, refinement, and function of carbonic anhydrase isozymes: refinement of human carbonic anhydrase I', *Annals of the New York Academy of Sciences*, vol. 429, pp. 49–60.

Kern, J., Biesiadka, J., Loll, B., Saenger, W. and Zouni, A. (2007) 'Structure of the Mn_4–Ca cluster as derived from X-ray diffraction', *Photosynthetic Research*, vol. 92, pp. 389–405.

Kiagawa, T. (2000) 'Structures of reduction intermediates of bovine cytochrome c oxidase probed by time-resolved vibrational spectroscopy', *Journal of Inorganic Biochemistry*, vol. 82, pp. 9–18.

Kitajima, N., Koda, T., Hashimoto, S., Kitagawa, T. and Moro-okaa, Y. (1988) 'An accurate synthetic model of oxyhaemocyanin', *Journal of the Chemical Society, Chemical Communications*, vol. 1988, pp. 151–2.

Kurtz, D.M., Shriver, D.F. and Klotz, I.M. (1976) 'Resonance Raman spectroscopy with unsymmetrically isotopic ligands – differentiation of possible structures of hemerythrin complexes', *Journal of the American Chemical Society*, vol. 98, pp. 5033–5.

Kurz, P., Berggren, G., Anderlund, M.A. and Styring, S. (2007) 'Oxygen evolving reactions catalysed by synthetic manganese complexes: A systematic screening', *Dalton Transactions*, vol. 2007, pp. 4258–61.

Lawson, D.M., Artymiuk, P.J., Yewdall, S.J., Smith, J.M., Livingstone, J.C., Treffry, A., Luzzago, A., Levi, S., Arosio, P., Cesareni, G., Thomas, C.D., Shaw, W.V. and Harrison, P.M. (1991) 'Solving the structure of human H ferritin by genetically engineering intermolecular crystal contacts', *Nature*, vol. 349, no. 6309, pp. 541–4.

Lee, H.-I., Cameron, L.M., Hales, B.J. and Hoffmann, B.M. (1997) 'CO binding to the FeMo-cofactor of CO-inhibited Nitrogenase: ^{13}CO and ^1H Q-Band ENDOR investigation', *Journal of the American Chemical Society,* vol. 119, pp. 10121–6.

Levine, S.A., Gordon, B. and Derick, C.L. (1924) 'Some changes in the chemical constituents of the blood following a marathon race', *Journal of the American Medical Association*, vol. 82, no. 22, p. 1779.

Licht, S.S., Lawrence, C.C. and Stubbs, J. (1999) 'Thermodynamic and kinetic studies on carbon–cobalt bond homolysis by ribonuclease triphosphotase reductase. The importance of entropy in catalysis', *Biochemistry*, vol. 38, pp. 1234–42.

Loll, B., Kern, J., Saenger, W., Zouni, A. and Biesiadka, J. (2005) 'Towards complete cofactor arrangement in the 3.0Å resolution structure of photosystem II', *Nature*, vol. 438, pp. 1040–5.

MacKinnon, R. (2003) Nobel Lecture 'Potassium Channels and the Atomic Basis of Selective Ion Conduction', available from http://nobelprize.org/nobel_prizes/chemistry/laureates/2003/mackinnon-lecture.html

Magnus, K.A., Hazes, B., Tonthat, H., Bonaventura, C., Bonaventuraand, J. and Hol, W.G.J. (1994) 'Crystallographic analysis of oxygenated and deoxygenated states of arthropod hemocyanin shows unusual differences', *Proteins–Structure, Function and Genetics,* vol. 19, pp. 302–9.

Mizutani, K., Yamashita, H., Kurokawa, H., Mikami, B. and Hirose, M. (1999) 'Alternative structural state of transferrin. The crystallographic analysis of iron-loaded but domain-opened ovotransferrin N-lobe', *Journal of Biological Chemistry*, vol. 274, pp. 10190–4.

Morth, J.P., Pedersen, B.P., Toustrup-Jensen, M.S., Sorensen, T.L.M., Petersen, J., Andersen, J.P., Vilsen, B. and Nissen, P. (2007) 'Crystal structure of the sodium–potassium pump', *Nature*, vol. 450, p. 1043–U6.

Muramoto, K., Hirata, K., Shinzawa-Itoh, K., Yoko-o, S., Yamashita, E., Aoyama, H., Tsukihara, T. and Yoshikawa, S. (2007) 'A histidine residue acting as a controlling site for dioxygen reduction and proton pumping by cytochrome c oxidase', *Proceedings of the National Academy of Sciences of the USA*, vol. 104, no. 19, pp. 7881–6.

Pan, Y-H., Sader, K., Powell, J.J., Bleloch, A., Gass, M., Trinick, J., Warley, A., Li, A., Brydson, R. and Brown, A. (2009) '3D morphology of the human hepatic ferritin mineral core: New evidence for a subunit structure revealed by single particle analysis of HAADF-STEM images', *Journal of Structural Biology*, vol. 166, pp. 22–31.

Pawelek, P.D., Croteau, N., Ng-Thow-Hing, C., Khursigara, C.M., Moiseeva, N., Allaire, M. and Coulton, J.W. (2006) 'Structure of TonB in complex with FhuA, *E. coli* outer membrane receptor', *Science*, vol. 312, pp. 1399–402.

Pelletier, H. and Kraut, J. (1992) 'Crystal structure of a complex between electron transfer partners, cytochrome c peroxidase and cytochrome c', *Science*, vol. 258, no. 5089, pp. 1748–55.

Peters, J.W., Stowell, M.H., Soltis, S.M., Finnegan, M.G., Johnson, M.K. and Rees, D.C. (1997) 'Redox-dependent structural changes in the nitrogenase P-cluster', *Biochemistry*, vol. 36, no. 6, pp. 1181–7.

Shaw, J.A., Macey, D.J. and Brooker, L.R. (2008) 'Radula synthesis by three species of iron mineralising molluscs: production rate and elemental demand', *Journal of the Marine Biological Society of the United Kingdom*, vol. 88, pp. 597–601.

Strop, P., Takahara, P.M., Chiu, H., Angove, H.C., Burgess, B.K. and Rees, D.C. (2001) 'Crystal structure of the all-ferrous [4Fe–4S](0) form of the nitrogenase iron protein from *Azotobacter vinelandii*', *Biochemistry*, vol. 40, no. 3, pp. 651–6.

Thamann, T.J., Loehr, I.S. and Loehr, T.M. (1977) 'Resonance Raman study of oxyhaemocyanin with unsymmetrically labelled oxygen', *Journal of the American Chemical Society*, vol. 99, p. 4188.

Tsukihara, T., Aoyama, H., Yamashita, E., Tomizaki, T., Yamaguchi, H., Shinzawa-Itoh, K., Nakashima, R., Yaono, R. and Yoshikawa, S. (1995) 'The whole structure of the 13-subunit oxidised cytochrome c oxidase at 2.8Å', *Science*, vol. 269, pp. 1069–74.

Yang, A.H., MacGillivray, R.T., Chen, J., Luo, Y., Wang, Y., Brayer, G.D., Mason, A.B., Woodworth, R.C. and Murphy, M.E. (2000) 'Crystal structures of two mutants (K206Q, H207E) of the N-lobe of human transferrin with increased affinity for iron', *Protein Science*, vol. 9, pp. 49–52.

Yano, J., Kern, J., Irrgang, K.D., Latimer, M.J., Bergmann, U., Glatzel, P., Pushkar, Y., Biesiadka, J., Loll, B., Sauer, K., Messinger, J., Zouni, A. and Yachandra, V.K. (2005) 'X-ray damage to the Mn_4Ca complex in single crystals of photosystem II: a case study for metalloprotein crystallography', *Proceedings of the National Academy of Science*, vol. 102, pp. 12047–52.

Acknowledgements

Grateful acknowledgement is made to the following sources for permission to reproduce material in this text.

Cover and title page

Computer model of the molecular structure of the protein cytochrome c550. Phantatomix/ Science Photo Library.

Figures

Figure 1.2b: Adapted from Bertini, I. et al. (2007) *Biological Inorganic Chemistry*, University Science Books; Figure 1.3: Frausto da Silva, J.J.R. and Williams, R.J.P. (2001) *The Biological Chemistry of the Elements-The inorganic chemistry of life*, 2nd edn, Oxford University Press; Figure 4.9: Doyle, D.A. et al. (1998) 'The structure of the potassium channel: molecular basis of K+ conduction and selectivity', *Science*, 3 April 1998. American Association for the Advancement of Science; Figure 4.17: Gadsby, D.C. (2007) 'Ion pumps made crystal clear', *Nature*, 13 December 2007. Nature Publishing Group; Figure 4.18a: Biophoto Associates; Figure 4.22: Nick Hobgood, Wikipedia; Figure 4.25: Lippard, S.J. and Berg, J.M. (1994) *Principles of Bioinorganic Chemistry*, University Science Books; Figure 4.26: Adapted from Bertini, I. et al. (2007) *Biological Inorganic Chemistry*, University Science Books; Figure 5.2: St Pierre, T.G. et al. (1996) 'Synthesis, structure and magnetic properties of ferritin cores', *Coordination Chemistry Reviews*, 151. Elsevier Science; Figure 5.3: Bertini, I. et al. (2007) *Biological Inorganic Chemistry*, University Science Books; Figure 6.19: Pan, Y. H. et al. (2009) *Journal of Structural Biology*, vol. 166, 2009. Elsevier Science; Figures 6.20 and 6.22: Adapted from Mann, S. (2001) *Biomineralisation: Principles and Concepts in Bioinorganic Materials Chemistry*, Oxford University Press; Figure 6.21: Heather Angel/Natural Visions; Figures 7.8, 7.9, 7.13 and 7.15: Adapted from Bertini, I. et al. (2007) *Biological Inorganic Chemistry*, University Science Books; Figure 8.16: Kurtz, D.M. et al. (1976) 'Resonance Raman spectroscopy with unsymmetrically isotopic ligands', *Journal of the American Chemical Society*, 98 (16). Copyright © 1976 American Chemical Society; Figure 8.19: Thamann, T.J., Loehr, I.S., and Loehr, T.M. (1977) 'Resonance Raman study of oxyhaemocyanin with unsymmetrically labelled oxygen', *Journal of the American Chemical Society*, 99. Copyright © 1977 American Chemical Society; Figure 8.21: Kitajima, N. et al. (1988) 'An accurate synthetic model of oxyhaemocyanin', *Journal of the Chemical Society, Chemical Communications*, 2, Copyright © 1988 American Chemical Society; Figure 8.30: Daresbury Laboratory, Cheshire; Figure 8.34: Adapted from Bertini, I. et al. (2007) *Biological Inorganic Chemistry*, University Science Books.

A number of the figures in this publication were created using data from The Protein databank http://www.rcsb.org.

We wish to acknowledge the use of the Chemical Database Service at Daresbury. The United Kingdom Chemical Database Service. Fletcher, D.A., McMeeking, R.F., Parkin, D., *J. Chem. Inf. Comput. Sci.* 1996, vol. 36, pp. 746–749 and the Inorganic Crystal Structure Database (ICSD). Bergerhoff, G. and Brown, I.D. in *Crystallographic Databases*, F.H. Allen et al. (Hrsg.) Chester, International Union of Crystallography (1987).

Every effort has been made to contact copyright holders. If any have been inadvertently overlooked the publishers will be pleased to make the necessary arrangements at the first opportunity.

Index

Entries and page numbers in **bold type** refer to key words which are printed in **bold** in the text. Indexed information on pages indicated by *italics* is carried mainly or wholly in a figure or a table. Entries present in Chapter 9 (the online chapter) are indicated by 'Ch9'.

RETURN TO: CHEMISTRY LIBRARY

100 Hildebrand Hall • 510-642-3753

LOAN PERIOD 1	2	3
4	5	6 1-MONTH USE

ALL BOOKS MAY BE RECALLED AFTER 7 DAYS

NOV 21 DUE AS STAMPED BELOW

NOV 21		

FORM NO. DD 10
1M   6-09

UNIVERSITY OF CALIFORNIA, BERKELEY
Berkeley, California 94720–6000